John A. D'Addario
Albert Einstein College of Medicine

EPISOMES

John A. D'Addario
Albert Einstein College of Medicine

MODERN PERSPECTIVES IN BIOLOGY

Under the Editorship of

HARLYN O. HALVORSON *The University of Wisconsin*
MOLECULAR BIOLOGY

HERSCHEL L. ROMAN *University of Washington*
GENETICS

EUGENE BELL *Massachusetts Institute of Technology*
DEVELOPMENTAL BIOLOGY

Published:

A HISTORY OF GENETICS
A. H. Sturtevant CALIFORNIA INSTITUTE OF TECHNOLOGY

MOLECULAR ORGANIZATION AND BIOLOGICAL FUNCTION
John M. Allen, Editor THE UNIVERSITY OF MICHIGAN

BIOSYNTHESIS OF SMALL MOLECULES
Georges N. Cohen—CENTRE DE LA RECHERCHE SCIENTIFIQUE, GIF-SUR-YVETTE, FRANCE

THE GENETIC CODE: The Molecular Basis for Genetic Expression
Carl R. Woese UNIVERSITY OF ILLINOIS

THE PRIMARY STRUCTURE OF PROTEINS: Principles and Practices for the Determination of Amino Acid Sequence
Walter A. Schroeder CALIFORNIA INSTITUTE OF TECHNOLOGY

EPISOMES
Allan M. Campbell STANFORD UNIVERSITY

ENZYMIC CATALYSIS
John Westley UNIVERSITY OF CHICAGO

Allan M. Campbell
Department of Biological Sciences
Stanford University
Stanford, California

EPISOMES

HARPER & ROW, PUBLISHERS • NEW YORK, EVANSTON, AND LONDON

EPISOMES

Copyright © 1969 by Allan M. Campbell

Printed in the United States of America. All rights reserved. No part of this book may be used or reproduced in any manner whatsoever without written permission except in the case of brief quotations embodied in critical articles and reviews. For information address Harper & Row, Publishers, Incorporated, 49 East 33rd Street, New York, N.Y. 10016.

LIBRARY OF CONGRESS CATALOG CARD NUMBER: 69-13746

To Alice, Wendy, and Joseph

CONTENTS

Preface	xi
Gene Symbols Used in the Text	xii
1 *Historical Introduction*	1
LYSOGENY	1
FERTILITY FACTORS	6
COLICINOGENY	11
THE EPISOME CONCEPT	13
2 *Temperate Bacteriophages*	15
COLIPHAGE LAMBDA	15
COLIPHAGE P1	25
COLIPHAGE P2	28
Salmonella PHAGE P22	29
OTHER TEMPERATE PHAGES	31
3 *Transfer Agents*	35
THE FERTILITY AGENT F OF *E. coli* K-12	35
COMPOSITE TRANSFER AGENTS	38
RESISTANCE TRANSFER AGENTS	40
THE R AGENT OF *Shigella flexneri* 2B STRAIN 222	40
THE Δ FACTOR	46
EPIDEMIOLOGY OF R AGENTS	51
COLICINOGENY AGENTS	53
4 *Bacterial Plasmids and Partial Diploidy*	57
ABERRANT INHERITANCE IN CROSSES	57
VEGETATIVE SEGREGATION	59
REPLICATION-DEFICIENT MUTANTS	66
STAPHYLOCOCCAL PLASMIDS	66

	PLAN FOR REMAINDER OF BOOK	67
5	*Mode of Chromosomal Attachment*	68
	MULTIPLE ATTACHMENT SITES AND ALTERNATIVE MODES OF ATTACHMENT	77
6	*Mechanism of Chromosomal Attachment*	81
	PHYSICAL NATURE OF THE RECOMBINATIONAL EVENT	81
	NATURE AND LOCATION OF THE CROSSOVER REGION	82
	PHYSICAL STUDIES OF BASE SEQUENCE SIMILARITIES	87
	STERIC HINDRANCE AND FACILITATION	88
	GENETIC FACTORS IN PROPHAGE INTEGRATION	92
7	*Mechanism of Detachment*	95
	CURING	96
	DETACHMENT OF THE F FACTOR	97
8	*Abnormal Detachment and the Formation of Transducing Phages*	99
	GENE PICKUP BY EPISOMES	99
	GENERALIZED TRANSDUCTION	108
	THE CRYPTIC PROPHAGE	110
	GENETIC CONTROL OF ABNORMAL DETACHMENT	112
9	*Immunity and Its Genetic Control*	114
	REPRESSION AND SUPERINFECTION IMMUNITY	114
	GENERAL REMARKS CONCERNING GENE REGULATION	115
	GENETIC CONTROL OF IMMUNITY	122
	DISSECTION OF THE IMMUNITY REGION INTO COMPONENT ELEMENTS	124
	REGULATOR MUTANTS	124
	GENETIC PROGRAM AND REGULATORY CIRCUITRY OF THE LAMBDA VIRUS	126
	OPERATOR MUTANTS	129
	TITRATION OF THE IMMUNITY	131
	RELATION OF ATTACHMENT TO IMMUNITY AND GENE FUNCTION	135
	IMMUNITY OF TRANSFER FACTORS	136
	CONCLUSION	138
10	*Autonomous Replication*	139
	SEGREGATION OF EPISOMES AT CELL DIVISION	143
	INTERFERENCE BETWEEN EPISOMES	144

	CHROMOSOME REPLICATION AND BACTERIAL MATING	146
	RELATION BETWEEN MODE OF REPLICATION AND STATE OF ATTACHMENT	147
	INTERMEDIATE STRUCTURES IN REPLICATION	148
11 *Joining and Separation of Ends*		149
12 *Polylysogeny*		154
	PROPHAGE LOSS FROM DOUBLE LYSOGENS	154
	PHAGE PRODUCTION FROM DOUBLE LYSOGENS	159
	ADEQUACY OF THE THEORY	160
13 *Toward a Definition of the Episome*		161
14 *Episomes as Model Systems*		165
	EPISOMES AND DEVELOPMENT	165
	TUMOR VIRUSES	168
	EPISOMES AND CHROMOSOMAL MECHANICS	170
References		173
Index of Names		187
Subject Index		190

PREFACE

In 1962, I wrote a review on "Episomes" that appeared in *Advances in Genetics*. Since that time, much progress has been made in molecular biology and bacterial genetics. In this book I have reviewed the subject of episomes in light of this newer information.

An inevitable shortcoming of books about rapidly moving fields is that new facts are accumulating while writing is in progress. This volume was composed in the period 1966–1968. It has been necessary to balance the desire to incorporate new and important findings against the preservation of logical organization and coherence. My judgment as to which facts to append and which to ignore will probably not agree entirely with that of others.

I have elected to include in the book a certain amount of "historical" information, particularly in Chapters 1, 2, and 5. My intention, especially in Chapter 5, was to trace the development of the major concepts in the area. Chapter 5 is perhaps historically incomplete because it fails to note the strong influence that the ideas of classical cytogenetics played in the original formulation of my own ideas about lambda prophage attachment. They can be considered a part of "molecular biology," but they depend more on the known properties of maize chromosomes than on those of DNA molecules.

It is a pleasure to take this opportunity to thank several people who read the first version of the manuscript critically—especially William Dove, Robert Rownd, and Hamilton Smith. Several sections of the book have been substantially rewritten in light of their very pertinent comments. I am of course solely responsible for all errors and omissions that remain.

This book was written while I was in the Biology Department of the University of Rochester and the recipient of a Research Career Award from the National Institutes of Health. I wish to thank both of these institutions for their support.

Stanford, California　　　　　　　　　　　Allan M. Campbell
October 21, 1968

GENE SYMBOLS USED IN THE TEXT

Escherichia coli

ara:	inability to metabolize arabinose
arg:	requirement for arginine
aro:	requirement for aromatic amino acids
att80:	attachment site for prophage 80
attλ:	attachment site for prophage lambda
azi:	azide resistance
bio:	requirement for biotin
cam:	chloramphenicol resistance
chl:	chlorate resistance
col:	ability to produce colicin
cys:	requirement for cysteine
dsd:	inability to make D-serine deaminase
gal:	inability to ferment galactose
gua:	requirement for guanine
his:	requirement for histidine
lacI:	regulator for lac region
lacY:	inability to form β-galactoside permease
lacZ:	inability to form β-galactosidase
leu:	requirement for leucine
lys:	requirement for lysine
mal:	inability to hydrolyze maltose
met:	requirement for methionine
pho:	inability to form alkaline phosphatase
pro:	requirement for proline
pur:	requirement for purines
rec:	deficiency in recombination
ser:	requirement for serine
str:	streptomycin resistance
sul:	sulfonamide resistance
supB,C:	ochre suppressors

tet:	tetracycline resistance
thr:	threonine resistance
thy:	requirement for thymine
tof:	resistance to phages T1, T5
ton:	resistance to phage T1
trp:	requirement for tryptophan
tsx:	resistance to phage T6
valS:	resistance to valine
xyl:	inability to metabolize xylose

Phage lambda

att:	site of attachment to the bacterial chromosome
b2:	low buoyant density because of DNA deletion, structural inability to attach to bacterial chromosome
c:	formation of clear plaques, due to absence of lysogenic response
h:	ability to adsorb to bacterial mutants resistant to wild-type phage
hs:	inviability at high temperature
imm:	locus determining immunity specificity
ind:	noninducibility by ultraviolet light
int:	inability to elaborate physiological factor(s) necessary for prophage integration
m:	medium-sized plaques, smaller than wild type
mi:	minute plaques, frequently with halos
rex:	unable to exclude *rII* mutants of phage T4
ri:	insensitive to replication inhibition
sus:	dependence on suppressor genes of the host
t:	defective (unable to form plaques)
v:	virulent, able to make plaques on lysogenic indicators

1

HISTORICAL INTRODUCTION

LYSOGENY

In the early 1920s, shortly following the discovery of bacterial viruses themselves, it was found that certain bacterial strains were lysogenic: Cultures of these strains, not recently phage infected, nonetheless produced phage. Their phage-generating ability was not destroyed by standard methods for purification of bacterial cultures from adventitious contaminants. During the subsequent three decades, the phenomenon remained more or less in limbo. Some very penetrating surmises were made, but the experiments left room for doubt as to whether the basic observations were not attributable to trivial technical artifacts. These doubts were finally dispelled around 1950 by the work of Lwoff and his associates at the Institut Pasteur.

What Lwoff showed was that the state of being lysogenic is passed from a bacterium to its descendants under conditions where external reinfection of individual cells by free phage particles in the surrounding medium had been made virtually impossible. From this he concluded that lysogeny, in the sense it had appeared to earlier workers, i.e., the intracellular reproduction and transmission of the capacity to produce phage, really did exist. It became necessary to postulate the existence of a specific genetic element (prophage) that determines this phage-producing ability.

Two further elementary facts about lysogeny were known: (1) Lysogeny can be acquired by infection. Many bacteria are already lysogenic as isolated from nature, but artificial lysogens can be created by phage infection of nonlysogenic bacteria. (2) Lysogeny is specific. Each natural lysogen produces a single type, or a limited number of types, of phage particles. Artificial lysogens produce the same kind of phage that was used for lysogenization. This specificity is found not only for different species of phage but also for mutant variants of a particular species. For example, if a nonlysogenic bacterium is infected with the *h sus-5* mutant of bacteriophage lambda, the lysogenic descendants of this bacterium liberate phage that are lambda *h sus-5* and no other type.

This proves that the prophage contains all the genetic information of the phage itself. However, artificial lysis of lysogenic cells reveals no infectious phage within them. Prophage is a latent, non-infectious form of the virus. Phage types which are able to establish lysogenic systems and to reproduce as prophage are called *temperate* phages, as distinguished from *virulent* phages which are unable to do so.

The existence of lysogeny raises two problems which, at various levels, will concern us through much of this book:

1. *The physiological problem.* The prophage is a virus genome. It is potentially able to take over and to destroy the bacterial cell. Occasionally it does so. That is why phage are found in lysogenic cultures, and that is how we know that lysogeny exists. The lysogenic cell thus harbors an element potentially capable of destroying it, and in order for the whole complex to persist as it does, the activities of the element must be repressed or controlled.

2. *The genetic problem.* Lysogeny behaves as a stable, hereditary property of a bacterial strain. Lysogeny denotes the presence of prophage. When a lysogenic cell divides, each daughter cell is lysogenic and therefore contains a prophage. How is prophage replication and segregation ordered so as to accomplish this end?

The physiological problem will be discussed in Chap. 9. The genetic problem was solved at a certain level by bacterial crosses among lysogenic strains.

For some phages, bacterial crosses showed that the prophage behaved like a bacterial gene. It could be "mapped" at a particular locus on the bacterial chromosome like any other marker. The chromosomal locations of some prophages of *Escherichia coli* are shown in Fig. 1-1. The most definitive evidence on prophage location comes from crosses in which each bacterial parent is lysogenic for a different variant of the same phage type; for example, between *gal+ λh bio* and *gal λh+ bio+*. Crosses between one lysogenic and one nonlysogenic parent give concordant results, but they are inherently less informative. Such crosses, in principle, cannot tell us anything about prophage location. The best they can possibly do is to show that some factor necessary for prophage growth or maintenance is at the chromosomal site. Whether this factor is the prophage itself (defined as the structure that bears the genetic information of the phage) can only be settled by an experiment that asks where such genetic information resides.

This genetic finding shows that a copy of the phage genome is closely bound to a specific site on the bacterial chromosome and replicates at least as often as the bacterial chromosome itself. When the bacterial chromosome divides, at least one copy of the phage genome

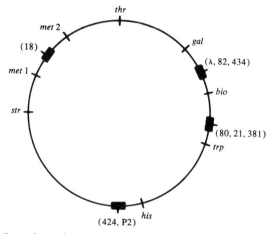

FIG. 1-1. Location of some prophages on the genetic map of *E. coli* K-12. Where several prophages are shown at the same site, the first one listed has been accurately located at the indicated site. The others in the same group might attach at the same site or at nearby chromosomal sites; some of the complications in distinguishing these possibilities are mentioned in Chap. 6. The location of lambda between *gal* and *bio* comes from Rothman (1965). Phage 80 was mapped on the *gal* side of *trp* by Signer et al. (1965). The locations of phages 424 and 18 are taken from Jacob and Wollman (1961), as are the approximate locations of phages 82, 434, 21, and 381. A P2 attachment site near *his* was demonstrated by Sunshine and Kelly (1968).

must be attached to each daughter chromosome. Further, if any extrachromosomal copies of the phage are in the lysogenic cell, these must not be perpetuated in separate, independent lines of descent. If such perpetuation occurred, introduction of another genetically marked phage genome by crossing (or by superinfection) would generally produce mixed yielders rather than pure lysogenic types. The rarity of mixed yielders shows that this is not the case. These rare exceptions turn out to be cases in which more than one phage genome has become attached to the chromosome. Indeed, as early as 1953, it was argued purely from the results of superinfection experiments, independently of bacterial crosses, that the lysogenic cell contained a very small number (probably one) of prophages. The prophage retained a special status and did not become mixed in a pool with the genomes of superinfecting phages. It is therefore reasonable that we give the name *prophage* to the chromosomally located phage genome, because genetically it is the sole carrier of phage information from ancestor to descendant in the lysogenic culture.

From a formal viewpoint, these results leave open the possibility that the lysogenic cell might generally contain copies of the phage genome other than the one master copy we call prophage. Only with the advent of DNA hybridization studies has it become directly demonstrable that the number of copies of the phage genome in a lysogenic cell is small, on the order of one per nucleus.

When a population of bacteria is infected with a temperate phage, some of the cells lyse and liberate progeny phage particles, whereas others survive to engender lysogenic progeny. Of those cells that lyse and liberate phage, the time elapsing between infection and lysis is typically of the same order of magnitude as one bacterial doubling cycle. A single infecting particle may, during this period, engender several hundred progeny. Obviously under these circumstances, the phage genome is replicating much more rapidly than it does as a prophage. This rapid replication culminating in phage production and lysis is termed *autonomous* (or sometimes, for historical reasons, vegetative) *multiplication*. The phage genome is in the *autonomous state,* as contrasted with the *integrated state* of the prophage.

Now, clearly, we should not wish to distinguish two states of replication merely by a quantitative difference in rates. Such a difference may reflect a qualitative difference, but we would like to define the nature of that qualitative difference as rigorously as possible. The idea naturally suggested by the facts is that the prophage multiplies as part of the bacterial chromosome, whereas autonomous phage multiplies independently. The idea is in fact so natural that it has been frequently taken for granted, and a detailed discussion may seem to belabor the obvious. It nevertheless seems important to state the case explicitly, because the notion of autonomous replication has been applied not only to phage but to many other objects for which the operational definition is much less direct.

The key point is what we mean, operationally, by "independent." Phage multiplication is always indirectly dependent on the bacterial chromosome because the products of that chromosome's action constitute much of the cellular machinery that the phage uses for growth. What we would like to show is that autonomous replication is less dependent on the chromosome than is prophage replication.

The difference between vegetative multiplication and prophage multiplication is determined at the cellular level. Either a cell lyses and produces phage or else it survives and becomes lysogenic. The two types of replication generally do not take place in the same cell at the same time because of the determinative role played in the process by diffusible physiological factors, which we shall discuss later (Chaps. 9 and 10). Given that the fundamental determination is at the cellular level, we could imagine that there is really no physical difference between autonomous phage and prophage, only between the cells in which they happen to find themselves. Two pertinent questions can be asked:

1. Since a prophage divides when and only when the bacterial chromosome divides, could the bacterial chromosome divide every time autonomous phage does? Certainly not. The total amount of DNA synthesis required would be impossibly high.

2. Could autonomous phage be associated physically with the bacterial chromosome while it multiplies? There is little direct evidence. Some virulent phages cause a virtually complete degradation of bacterial DNA; so such association is not required for phage multiplication in general. For temperate phages, these facts seem pertinent: Mutant phages unable to attach to the chromosome and become prophages can multiply autonomously. Likewise, wild-type phage can grow in bacterial strains from which the site of prophage attachment has been deleted. Among the phage produced from lysogenic cells are found occasional particles that have picked up contiguous chromosomal genes (specialized transduction, see Chap. 8). Such specialized transduction is not observed with lytically grown phage, even when the phage is genetically able to attach to the chromosome. Both observations suggest that the physical relationship, if any, between autonomous phage and bacterial chromosome differs from that between prophage and chromosome.

We conclude that a phage genome can indeed multiply in two different states: integrated, as a prophage, or autonomous. Conceptually, the distinction between autonomous and integrated has two separate aspects: (1) the nature of replication control and (2) the physical relationship between prophage and bacterial chromosome. The connection between autonomous replication and physical detachment will be treated in more detail in Chap. 10.

FERTILITY FACTORS

During the early 1950s, while Lwoff was busy bringing lysogeny into the realm of respectable science, the determination of bacterial fertility was under study in several laboratories, especially in Wisconsin by the Lederbergs and in London by Hayes. It had been shown earlier by Lederberg and Tatum that mutants of *Escherichia coli* K-12 could recombine genetically with each other. The experiment was to mix the two mutant strains on the agar surface of a medium that allowed growth of neither mutant but only of the parent strain from which both were derived. The result indicated not only the existence of recombination but also the absence of any sexual differentiation, as any two mutant strains would mate.

Several years and many crosses later, it was discovered that this is not invariably so. There are pairs of strains that fail to produce recombinants when crossed with each other, although both are fertile with most other strains. Tabulation of the results of all pairwise combinations classified strains into several types, two of which predominated: always fertile (F^+), which produce recombinants when crossed with either F^+ or F^-; and sometimes infertile (F^-), which yield no recombinants when crossed with other F^- strains. It was also observed that $F^+ \times F^-$ crosses produced many more recombinants than $F^+ \times F^+$ crosses did.

Since fertility is a hereditary property of the strain, there must be a genetic difference between F^+ and F^- stocks. Like any other hereditary character, this difference can be analyzed genetically from the results of crosses, even though the character itself is what made crossing possible in the first place. From these crosses, the following facts emerged:[1]

1. All the recombinants from $F^+ \times F^-$ crosses are F^+. The character therefore cannot be assigned a specific locus on the bacterial chromosome.

2. Fertility is infectious. When F^+ and F^- cultures are mixed, the F^- cells are converted to F^+. The rate of conversion is much higher than the rate of recombination.

3. Various treatments (exposure to acridine dyes or cobalt ion, thymine starvation, X-irradiation and others) convert some fraction of F^+ cells into F^-. The conversion is irreversible. No known treatment (other than fresh contact with F^+ cells) will change F^- cells back to F^+.

[1] For references and experimental details, see Chap. 4 of Jacob and Wollman (1961).

From these facts, it was surmised that the difference between F^+ and F^- cells was determined by an independent genetic element (fertility, or F, factor) present in F^+ strains but not in F^- ones. Unlike bacteriophages, which can be identified as physical particles in the free infectious phase of their life cycle, the fertility factors are agents whose existence was inferred purely from formal genetic analysis of bacterial crosses. Only much later, by coupling some genetic tricks with modern physical techniques, Falkow and Citarella (1965) could demonstrate a specific "F DNA." Thus the genetic particle can at last begin to be identified with a physical counterpart.

In 1950, Cavalli found a mutant of an F^+ *E. coli* which, when mated to an F^- strain, generated an unusually large number of recombinants. Such strains, of which many other examples were found later, were designated Hfr (for "high frequency of recombination"). $Hfr \times F^-$ crosses were analyzed by Hayes, and later by Jacob and Wollman.

Using the Hayes Hfr strain (HfrH), Jacob and Wollman showed that, in an $Hfr \times F^-$ cross, there is an oriented and fairly uniform transfer of the bacterial chromosome from the Hfr to the F^- cell. The entire bacterial chromosome could be mapped by measuring, for each marker, the minimum time required for transfer from Hfr donor to F^- recipient. The time of transfer was defined, operationally, as the time at which recombinant formation can no longer be prevented by subjecting the cells to shearing forces adequate to break apart mating pairs. This mapping procedure correlated well with the results of linkage studies.

Of special interest was the mapping of the Hfr character itself. The time of transfer of this character is later than that of any known gene. Coupled with linkage studies, this showed that the Hfr property maps at one terminus of the bacterial chromosome.

In 1956, Skaar and Garen also analyzed chromosome transfer from Hfr to F^- cells, using the original Hfr strain of Cavalli (Hfr C). Hfr C, like Hfr H, showed an oriented transfer of markers. However, the order of transfer was different from that found for Hfr H. Whereas Jacob and Wollman had observed the order

thr leu azi ton lac gal[2]

Skaar and Garen found instead

lac ton azi leu thr gal

The two orders are different but not entirely independent. They start at different places and go in opposite directions, but otherwise

[2] For definitions of genetic symbols, see p. xiii.

they are the same. If we imagine that the genes of the F^+ progenitor were arranged on a closed (circular) structure, the two Hfr derivatives are distinguished by different origins and opposite orientations on this circle (Fig. 1-2).

Jacob and Wollman isolated a collection of Hfr derivatives of *E. coli* K-12 and showed that the same pattern fit not just the Hfr strains of Hayes and Cavalli but all subsequent isolates as well. All isolates behave as though derived from the same circular linkage map,

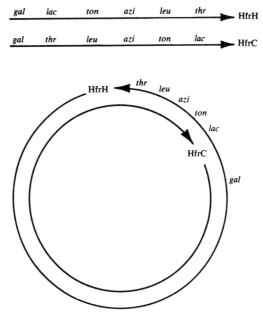

FIG. 1-2. Diagrammatic representation of the order of gene transfer by HfrH and HfrC. These two orders, taken together, generate a circular linkage map.

i.e., the genetic maps of the individual Hfr's comprise a collection of linear orders that are circular permutations and reverse circular permutations of each other.

As mentioned above, the "Hfr" determinant is located at the terminus of the linkage map. This is true for each of several isolates examined. The implication is that, in the formation of each Hfr strain, the chromosome has undergone some hereditary alteration at the site of origin of gene transfer from that strain.

What is the nature of this alteration? Good reasons soon developed for believing that it comprised the attachment of the F factor to the chromosome (at different chromosomal sites in different Hfr strains). One reason for thinking this was that Hfr strains occasionally revert to the F^+ state. This showed that the Hfr strain could not have lost the F factor.

Furthermore, there was no evidence that Hfr strains harbored any F agent in its usual state. Whereas the F^+ character is highly and rapidly infectious in mixed culture, Hfr strains do not convert F^- to F^+. Most of the recombinants from an $Hfr \times F^-$ cross are F^-, and those few that receive the fertility determinant are Hfr. It seemed as though normal "unattached" F was absent from the Hfr strain and reappeared only when it reverted to F^+, at which time the factor came off the chromosome. The alternative that the F agent remains present but unexpressed in the cytoplasm of the Hfr culture (and that the chromosomal alteration determining the Hfr character is something not involving F directly) is much harder to disprove than the analogous possibility with phage. It is now possible to label F genetically, so that more detailed studies of its behavior in $Hfr \times Hfr$ crosses are possible. We shall see later (Chap. 8) that the origin of certain variants of F is easier to understand if F itself is integrated.

So the F factor shares with the phage genome the possibility of replicating in two distinct states—autonomous and integrated. In the autonomous state, it is highly infectious to F^- cells growing in the same culture, and is curable by acridine dyes and other agents. The frequency of transfer of bacterial genes to other cells is low; in fact, much of the apparent donor ability of F^+ cells in crosses stems from the presence of Hfr variants in the F^+ culture. Integrated F is noninfectious and incurable, but promotes high frequency transfer of the entire bacterial chromosome.

As with phage, we must assume an interference at the cellular level between these two states—autonomous and integrated. Hfr strains generally do not allow autonomous multiplication of free F in the same cell, for reasons not thoroughly understood (cf. Chap. 10). Because of the indirect nature of the arguments, there is no rigorous proof that the state of the element actually differs, rather than just the state of the cell. It is easiest to imagine autonomous F as physically unconnected to the bacterial chromosome, but this remains to be demonstrated.

In the case of phage, autonomous replication is much faster than prophage replication and culminates in cellular destruction. Autonomous replication of F, on the other hand, continues indefinitely, and therefore the steady growth rate must necessarily be the same as that of the cell in which it grows. That autonomous replication of F is potentially faster than cellular growth can be seen in a mixed culture where F is infecting fresh F^- cells. The fraction of F^+ cells continually increases, showing that F is multiplying more rapidly than the cell. For example, consider a cell that contains one F particle. During a division cycle it transfers one F to another cell, as well as dividing to

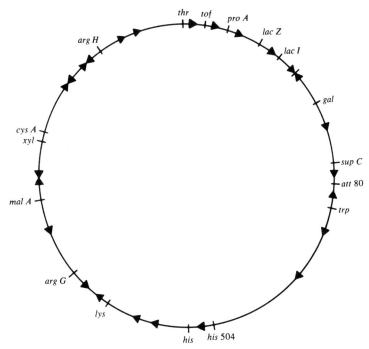

FIG. 1-3. Genetic map of *Escherichia coli* K-12, showing chromosomal sites of F attachment and direction of transfer by Hfr strains. Each different type of Hfr is symbolized by an arrowhead pointing in the direction of chromosome transfer; i.e., the gene behind the arrowhead is the first one transferred. Adapted from the map of Gross (1964), supplemented with information from Matney *et al.* (1964), Taylor and Thomas (1964), Curtiss (1964a), Signer and Beckwith (1966), Low (1967), and Taylor and Trotter (1967).

produce two F^+ daughter cells. The total number of F particles has increased by a factor of 3 during a time when the cell number doubled.

A map of *Escherichia coli* showing known loci of F integration is given in Fig. 1-3. More powerful selective techniques indicate that this is a small fraction of the total number of possible sites where integration can take place (cf. Chap. 5).

COLICINOGENY

In 1932, Gratia discovered that certain strains of *Escherichia coli* produced diffusible substances that are lethal to other strains of the same or related bacterial species. To these products was given the name "colicins," because they kill *Escherichia coli*. Analogous substances, termed in general "bacteriocins," are produced by various bacterial species. Colicinogeny resembles lysogeny insofar as bacteria with either property produce a specific agent lethal to certain other bacteria. The difference is that colicins are not phages and cannot multiply on the cells they kill. They contain no detectable nucleic acid, only protein and sometimes lipopolysaccharide. Like phages, colicins are highly specific. A variety of colicins exist, each having its own specificity of attachment to the bacterial surface and therefore its own range of sensitive hosts.

In the early 1950s, Fredericq and his co-workers made rather extensive studies on the assortment of the colicinogeny determinant (col) among recombinants from $F^+ \times F^-$ crosses in which one parent was col^+ and the other col^-. They were unable to map the col^+ determinant on the bacterial linkage map. Various colicins caused different complications. In the case of colicin El, all the progeny are col^+ when the F^- parent is col^+; and most recombinants are also col^+ in the reciprocal cross. This behavior is similar to that of the F agent, and suggested that the col character might likewise be determined by a nonchromosomal element.

In 1957, Alfoldi, Jacob, and Wollman studied this same colicinogeny determinant in $Hfr \times F^-$ crosses. They used several Hfr strains with different points of origin, and performed reciprocal crosses. Their results indicated that the *col* determinant was located at a specific site on the bacterial chromosome.

When an F^- cell acquires genetic markers by conjugation, it does not in general produce a pure clone of recombinant progeny. One reason is that integration of the new genes into the bacterial chromosome may be delayed. Also, *E. coli* cells commonly contain more than one chromosome. Integration will occur in only one of them, which is segregated from other chromosomes at later divisions. However, when an F^- col^- cell becomes col^+ by conjugation, all of its descendants are likewise col^+. This is similar to the behavior of F, and quite different from that of ordinary genetic markers (Fig. 1-4).

Alfoldi and co-workers thus concluded that colicinogeny determinants resembled the fertility factor and the genome of a temperate phage in the ability of all three to exist in two alternative states—auton-

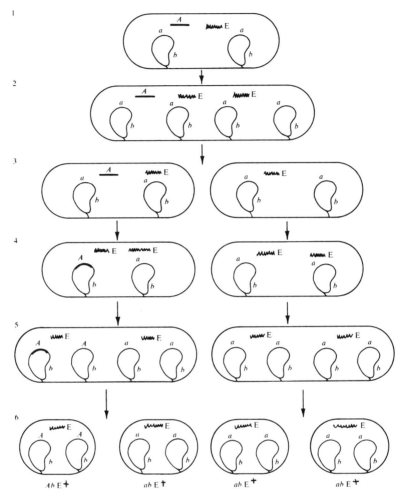

FIG. 1-4. Diagrammatic representation of the distribution of chromosomal genes and episomes in an originally F⁻ cell after conjugation. At stage 1, a cell is shown with two chromosomes carrying genetic markers *a* and *b*. This cell has just received by conjugation a fragment of the donor chromosome carrying the *A* allele of the *a* marker, and also an element E (such as F or *col*) that can replicate autonomously. At stage 2, both the chromosomes and E have replicated once. At stage 3, cell division has occurred. At stage 4, A has replaced its allele in one of the chromosomes, thus producing a recombinant chromosome (*Ab*). (The fate of the expelled *a* allele is not indicated.) At stage 5, another round of chromosomal replication has transpired. Another cell division is required to generate a genetically pure recombinant cell (stage 6).

Notice the orientation of chromosomal division with respect to cell division. The two daughters of the chromosome on the left at stage 1 pass omous and integrated. Subsequent work has failed to confirm the existence of an integrated state for the colicinogeny determinant that they studied; but, as we shall see in Chap. 3, other colicinogeny determinants can integrate into the chromosome.

THE EPISOME CONCEPT

In 1958, Jacob and Wollman published a brief note in *Comptes rendus* calling attention to the similarities of these elements and proposing the name "episome" for them. The most important features of episomes were the following: (1) An episome is an *added* element. The cell can survive without the episome. Episomes can be acquired externally, by infection or cellular conjugation. (2) An episome can reproduce itself in two alternative states—either autonomously or integrated into the bacterial chromosome. They later expanded their discussion of episomes and their possible relation to various biological problems and processes in a detailed review article and in a chapter of their book (1961).

Episomes are thus distinguished from chromosomal genes on the one hand and obligately cytoplasmic elements (plasmids) on the other. We may ask whether the mere existence of some objects able to exist in both states justifies the addition of another new word to the scientific language. Despite the fact that some generalizations concerning temperate phages have proven helpful in understanding fertility factors, for instance, it can be questioned whether episomes, as defined operationally, constitute a meaningful biological category.

We shall return to this theme in a more constructive vein (and in the light of facts detailed in intervening chapters) in Chap. 13. At this time, suffice it to say that it is useful to define a new category of objects if the very existence of such objects raises special questions that require

---

into the cell on the left at stage 3. The daughters of the chromosome on the right pass into the cell on the right. The chromosomes of stage 1 thus segregate from each other at cell division. Although the cell has two or four chromosomes at different stages, it is genetically haploid, because the two chromosomes of any one cell are always sisters. (The cytological diagram here is inferred from its genetic consequences rather than vice versa.)

The bacterial chromosome is drawn as circular and attached to the cell membrane (see Chap. 10). The autonomous element E is probably also circular and membrane-bound. It is represented here as linear and unattached for ease of presentation only. Notice that all the progeny of the exconjugant are E^+, although only some (¼ in this case) are *Ab*.

answers in terms of mechanism. And it is useful to define a category of objects if the existence of such objects provides a possible mechanism for known phenomena that require explanation. The main purpose of this book is to discuss the questions raised by episomes and, so far as we know them, their answers.

If a genetic element can become attached to a chromosome, the first question is how is it attached, i.e., what is the structure of the chromosome containing the integrated episome? At least in some cases, we can now make a fairly convincing case for what the mode of attachment is (Chaps. 5, 12). The second question is how it becomes attached to or detached from the chromosome. This is an area of active current research. Enough information is now available to assure us that this constitutes a real and interesting problem (Chaps. 6–8, 11). Third, we must ask how replication is controlled when the element is autonomous, and what happens to the control mechanism after integration. On this question there is little direct information available as yet.

A fourth question, less explicitly defined than the others, will be considered: What are the evolutionary implications of episomes—both for themselves and for their hosts? Here we shall mainly try to show that some of the information on the other questions has a real relevance to this problem (Chaps.13, 14).

Whereas the existence of episomes calls attention to the above questions, none is unique to episomes as strictly defined. For example, the questions of the mode and mechanism of attachment are equally raised by the transposable elements of maize, which are not known to multiply extrachromosomally. Similarly, regulation of autonomous replication is a problem equally applicable to obligate plasmids as to true episomes. In searching for answers, we shall pass freely from episomes to nonepisomes as the occasion demands.

2

TEMPERATE BACTERIOPHAGES

Lysogeny is very common among bacteria. Screening of bacterial isolates for ability to make plaques on related laboratory stocks reveals an array of phages. Because lysogeny frequently entails no major overt changes in colonial morphology or growth properties, potentiality to form phage can easily remain undetected.

For his studies on phage production in single cell pedigrees, Lwoff chose a lysogenic strain of *Bacillus megatherium* that had a high rate of spontaneous phage liberation. Later work has concentrated on a few temperate phages of the Enterobacteriacae—largely because of the volume of collateral information available on the genetics and biochemistry of the hosts.

Whereas the orientation of this book is primarily toward problems rather than material, it will clarify further discussion to give at this point a brief description of some common temperate phages and their properties. A history of the origin of laboratory stocks, more detailed than is customary, is included to allow the interested reader to evaluate the literature on these phages in a serious manner.

COLIPHAGE LAMBDA

About 1950, Esther Lederberg discovered that certain isolates of *Escherichia coli* K-12 which had survived heavy doses of ultraviolet irradiation were lysed by some product of the parent culture. This product proved to be a bacteriophage for which the parent strain was lysogenic. At that time, no information was available on prophage location, and it was a common assumption that prophage was extrachromosomal. Cytoplasmic factors have usually been designated by geneticists with Greek letters, and so the new phage was christened "lambda." Subsequently, the rationale has disappeared, but the name remains in use.

The original lambda phage was a rather poor object for genetic studies. The plaques were small, and differences in plaque morphology difficult to discern. In the early 1950s, mutants were isolated and genetic maps constructed by Jacob and Wollman and by Kaiser. Kaiser improved the stock by isolation of mutants that made better plaques than the original phage, and were therefore more suitable for genetic studies than the natural virus. The new improved phage was called "reference type" lambda by Kaiser in 1955. Reference type lambda had the undesirable property of not attaching stably to the bacterial chromosome. This defect was removed by subsequent outcrossing of Kaiser's stock with another lambda. The resulting phage, which forms good plaques and lysogenizes well, has been the object of most studies since 1957. It has been called "reference type lambda," or occasionally "lambda papa" (for *Pa*sadena and *Pa*ris, the source of the stocks used in its derivation). To avoid confusion, we shall refer to this stock as standard type lambda, reserving the name *reference type* for Kaiser's original 1955 stock.

Whereas this strain has become standard because of its convenient properties, both the original lambda and the original reference type still are used in some laboratories. Unless otherwise indicated, any general statements made here about lambda phage or lambda prophage apply to the standard type.

The first lambda mutants studied caused visible changes in plaque size or morphology. With these, it was possible to construct a linear genetic map, and to get some idea of the phage's mating behavior. In 1957, Jacob, Fuerst, and Wollman heralded a new era in phage genetics with a systematic study of defective mutants of lambda phage. They were not the first to observe defective lysogens, but no previous work of comparable scope had deliberately exploited the phenomenon to study the physiological genetics of phage.

Defective mutants are mutants that fail to go through a complete infectious cycle and therefore to make plaques. They survive because the genome of a temperate phage can multiply indefinitely in the prophage state without carrying on many of the functions required in the lytic cycle. The initial observation was that, from a mutagenized lysogenic culture, some bacteria can be recovered that are no longer lysogenic—but that are still immune to superinfection by phage of the type originally carried. Whereas such defective lysogens produce little or no phage, they contain phage genes that can be rescued by superinfection with various phage mutants. Superinfection results in the formation not only of active phage particles (superinfecting type and recombinants) but also, along with them, of defective particles. The defective particles do not form plaques, but they can lysogenize other bacteria and convert them to defective lysogens.

Jacob and co-workers saw in these observations a means of examining for the first time the entire functional genetic complement of a virus, or at least a major fraction thereof. It is reasonable to expect that most phage functions will be necessary for plaque formation, and unnecessary for prophage replication. Any mutation of the prophage that abolishes one or more such function will change the normal prophage into a defective one. By collecting defective mutants and characterizing the genetic block in each, one should be able to identify all the gene-controlled steps specific to phage reproduction. Each defective was accordingly tested for various phage-specific functions such as DNA replication, synthesis of particular proteins unique to the phage-infected cell, etc.

Work with defective mutants was limited for technical reasons. As they could be obtained as free particles only in mixed lysates, many types of experiments were difficult or impossible with them. This problem was circumvented by the study of conditional lethal mutants, which behave as defectives in some environments and as active phage in others. Heat-sensitive (*hs*) and suppressor-sensitive (*sus*) mutants have been widely used, especially the latter, which belong to the *amber* type, now well known from work on other systems.

With the *sus* mutants, it was easy to perform more extensive complementation studies and more accurate genetic mapping than were feasible with the true defectives. The mutants were classified into eighteen cistrons, which were assigned letters from A to R. The arrangement of cistrons along the genetic map was made as alphabetical as possible on the basis of the incomplete mapping data available at the time.

From physiological studies of conditional lethals as well as of true defectives, clustering of genes controlling related functions is apparent. This is illustrated in Fig. 2-1. Genes *A* to *F* are concerned with formation of phage heads. Mutants defective in these genes synthesize phage-specific DNA and functional tails, but fail to package the *DNA* into normal heads. Genes *G* to *J* function in tail formation. *J* is probably the structural gene for the protein that makes contact with the bacterial cell surface. In this gene are located mutations (*h*) affecting the host range of the phage particle by altering its adsorption properties (Siminovitch, 1967). Mutants lacking any or all of genes *G* to *J* make intact normal heads, which can be reconstituted to give infective particles by adding "tail donor" lysates from mutants of genes *A* to *F* (Weigle, 1966). More extensive mutant isolation has revealed one additional head gene (*W*) and three more tail genes (*ZUV*) (Parkinson, 1968).

After gene *J* comes a section of the map in which no phage-lethal mutations are found. This region functions in insertion of the phage genome into the bacterial chromosome. It contains the *b2* region, which plays a structural role in insertion, a locus (*att*) at which insertion actually takes place, and a gene *int* whose product plays a physiological role in the insertion process. (These elements will be discussed in detail in Chap. 6.) To the right of *int* are the structural genes for two phage-specific proteins (exonuclease and β; Radding *et al.*, 1967).

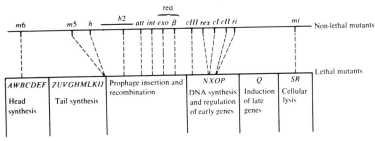

FIG. 2-1. Genetic map of phage lamba.

These two proteins seem to function in a recombinase system (*red*). Identification of some genetic components of the *red* system with the determinants for these proteins is under investigation at the time of writing.

Next comes a region of great complexity and interest, which controls repression of phage genes in the lysogenic cell. It also determines the early proteins which initiate phage reproduction and whose synthesis is directly repressed in the lysogenic cell. We do not know what any of these proteins may be, but there are three genes (*N*, *O*, and *P*), whose normal function is required for phage DNA synthesis (Joyner *et al.*, 1966). A group of early defective mutants (*X*) lie between *N* and *O* (Eisen *et al.*, 1966). Then there are three genes serving a regulatory function in lysogenization—not in prophage insertion, but rather in repression of those viral functions that otherwise would lead to cellular lysis. Mutants defective in these functions form plaques, but the plaques lack the overgrowth of lysogenic cells characteristic of wild-type (turbid) plaques and are instead clear (*c*).

Of these three genes (*cI*, *cII*, and *cIII*), only *cI* is directly concerned with repression in the lysogenic cell. Genes *cII* and *cIII* function during the period after infection when a cell decides whether or not to become lysogenic. They seem to be shut off after lysogeny becomes established (Bode and Kaiser, 1965).

Mutations (*rex*) affecting the ability of lambda prophage to exclude *rII* mutants of phage T4 map between *N* and *cI*. The *rex* function seems to be distinct from the *cI* function. Mutants insusceptible to direct replication inhibition by the immunity (*ri*) map between *cII* and *O* (Dove, 1968). They will be discussed in Chap. 10.

At the right end of the phage map come two genes, *Q* and *R*. Gene *Q* is not required for DNA synthesis, but is necessary for normal turn-on of late phage genes (Dove, 1966). *R* is the structural gene for the phage-specific lysozyme required for lysis (Campbell and del Campillo Campbell, 1963). Another gene (*S*) whose product is necessary for cellular lysis (but not for lysozyme production) maps close to *R*, probably to the left (Harris et al., 1967).

Lambda genetics has profited not only from studies of mutants, but also from the availability of related phages that could be crossed with lambda. Such interspecific crosses enable genetic localization of structural specificity, which is not possible when one is limited to mutants that have lost a given function rather than having evolved it in a different and complex way. In particular, it is possible by this means to study the specificity of prophage insertion in the bacterial chromosome and of superinfection immunity of lysogenic cells.

These two functions together probably tell us the whole story of lysogeny. Synthesis of heads, tails, early enzymes, etc., is a general property of bacterial viruses, temperate or not. As already discussed (Chap. 1), the existence of lysogeny poses two special problems that a temperate phage must solve—the physiological problem of repressing viral functions, and the genetic problem of assuring prophage transmission to daughter cells.

Genetic determination of insertion specificity and immunity specificity will be discussed in more detail in Chaps. 6 and 9, respectively. The two specificities are governed by distinct determinants, so that hybrids with the immunity of one phage (say, lambda) and the insertion specificity of another (e.g., phage 80) can be generated (Signer, 1964). Immunity is determined by the *cI* gene, whereas insertion specificity is determined by genes between *cIII* and *h*.

In other systems, it frequently has occurred that an apparently very simple relation between gene function and map position has been superseded by a more complex picture as new genes were discovered that did not fit the original scheme. It is unlikely that this will happen for phage lambda. The functional arrangement shown in Fig. 2-1 is consistent not only with the properties of point mutants in the individual genes but likewise with those of transducing variants having in

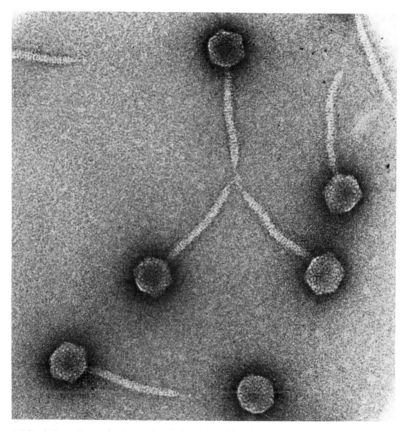

FIG. 2-2. Bacteriophage lambda. Magnification 180,000×. Uranyl acetate negative staining. (Electron micrograph kindly supplied by F. Eiserling.)

some cases extensive deletions of phage genes. Because the deleted region comprises a connected block, a function carried out by a deletion mutant must be independent of any genes (discovered or undiscovered) within the deleted region. Thus, the "tail" region of Fig. 2-1 contains no genes necessary for head formation (Campbell, 1967a);

and neither head nor tail region is required for phage DNA replication (Brooks, 1965; Joyner *et al.,* 1966).

Physically, the lambda particle weighs about 66×10^6 daltons, of which 33×10^6 daltons[1] comprise a single linear DNA molecule, 17.3μ in length (Caro, 1965). The only other known component is protein. The number of distinct protein species is unknown; from serological studies, at least four different antigens are inferred (Soller *et al.,* 1965). An isometric head of diameter 650 Å has a capsid apparently comprising 252 identical 70–80 Å subunits. The tail is 1530 Å long and 170 Å wide and shows regular striations at 44 Å intervals (Eiserling and Boy de la Tour, 1965).

The density of lambda is 1.51, whereas that of lambda DNA is 1.71. The DNA molecule is uninterrupted and double stranded throughout most of its length, but the ends of the molecule are single stranded for a region of about 20 nucleotides. The two ends are complementary to each other, and can anneal to produce either circular monomers or linear dimers. The terminal nucleotides on the two strands are known, as well as the approximate composition of the complementary region.

Lambda DNA molecules can be fractionated into parts by breakage in either the longitudinal or the transverse dimension. Shear degradation, which causes transverse splitting into double-stranded fragments, yields fractions of different densities, reflecting local differences in guanine+cytosine (GC) content of the DNA. Detailed study has revealed six regions of the lambda molecule, each with its characteristic GC content (Skalka, 1968).

Heating of lambda DNA causes the molecules to split longitudinally into single strands. These are separable by their different buoyant densities at alkaline pH, or in the presence of the artificial copolymer poly IG, which perhaps binds to clusters of cytosine residues (Szybalski *et al.,* 1966). That strand which is denser at alkaline pH is the one that is lighter in the presence of poly IG. By convention, it is called the "l" strand (because RNA transcription on it goes leftward on the genetic map as usually drawn). The "r" strand (transcribed rightward) is lighter at alkaline pH and denser in poly IG than the l strand (cf. Fig. 2-3).

Several methods have made it possible to equate the physical structure with the genetic map:

1. *Recombination studies.* When Meselson and Weigle (1961)

[1] One dalton is defined as one unit of atomic weight ($O_2 = 16$).

showed that genetic recombination of lambda is associated with physical chromosome breakage, they incidentally demonstrated that recombinants for the marker pair *c mi* receive most of their DNA from the parent contributing the *c* marker rather than the one contributing *mi*. This is the expected result if phage genes are linearly disposed on the DNA molecule, with *mi* close to one end and *c* nearer the molecular center. Extension of this method to other marker pairs can show the relation between recombination frequency and physical length (Jordan and Meselson, 1965). For technical reasons, this method is inapplicable to genes near the center of the molecule.

2. *Transformation studies.* Under appropriate conditions, lambda DNA can infect cells and engender phage progeny. One of the requisite conditions is simultaneous infection by a whole lambda particle

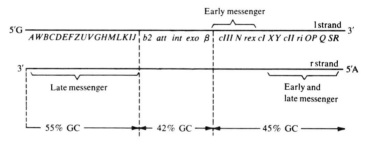

FIG. 2-3. Nomenclature and properties of the DNA strands of lambda phage. [Based on Szybalski, Kubinski, and Sheldrick (1966). Used by permission of the authors.]

(helper phage).[2] Phage resulting from DNA infection is distinguished from progeny of the helper phage itself by genetic markers. Marker rescue is observed not only with complete lambda DNA molecules but also with double-stranded fragments produced by shear degradation. The transformation assay is diagrammed in Fig. 2-4.

The only fragments with transforming activity are those that contain one end or the other of the complete molecule. Pieces coming from the center of the molecule are inactive. The markers most resistant to shear degradation are those closest to the ends of the molecule.

Where transforming DNA and helper phage differ by several markers, infection with whole DNA molecules usually leads to production of phage particles with the entire complement of phage markers

---

[2] Productive infection of protoplasts in the absence of helper has recently been achieved (Young and Sinsheimer, 1967).

from the DNA. With sheared DNA on the other hand, linkages are disrupted. A preparation of (on the average) half molecules shows virtually no linkage between markers at opposite ends of the genetic map.

From the dependence of single-marker and linked transformation on average fragment size, gene order along the DNA molecule can be

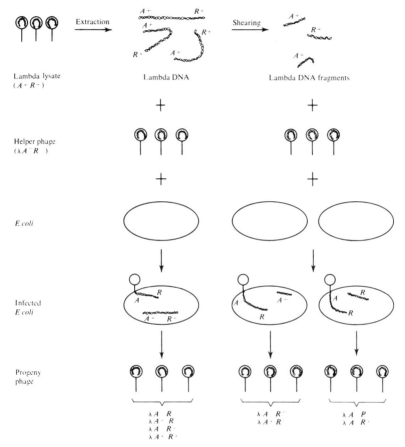

FIG. 2-4. Diagram of transformation assay with lambda DNA, as developed by Kaiser and Hogness.

determined. This order agrees with the genetic map based on ordinary crosses (Kaiser, 1962; Hogness et al., 1966).

3. *Density measurements of transducing phages.* Another method for estimating physical distances between genetic markers has been to measure the DNA content of transducing phages carrying different amounts of genetic material (Kayajanian and Campbell, 1966).

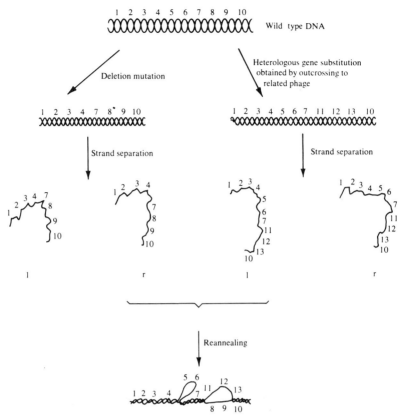

FIG. 2-5. Diagrammatic representation of the application of molecular cytogenetics to bacteriophage DNA. The gene order on the wild-type molecule is indicated on the top line. The second line shows a deletion mutant (such as lambda *b2*) and a phage with a heterologous gene substitution (such as lambda *imm434*). When the l strand of the deletion mutant is annealed to the r strand of the substituted phage, a single-stranded loop is formed at the site of the deletion. Heterologous substitution produces a pair of single-stranded regions, which may be of equal or unequal length depending on the particular substitution.

The discrimination of DNA coming from different sections of the lambda molecule has been made possible by the transformation work, but also by the availability of transducing phages, which have deletions of one section or another of the lambda map. (The origin of these abnormal particles will be described in Chap. 8). It is, for example, possible to examine the DNA or messenger RNA corresponding to the region from N to P by looking at that fraction which will hybridize with DNA from a biotin-transducing phage containing this region, but not with one lacking it. Deletions are now available dissecting the phage map in almost any desired way.

4. *Molecular cytogenetics.* Like other DNA molecules, lambda DNA will separate into single strands on heating and reanneal into double strands with slow cooling. If lambda DNA is mixed with DNA from a variant such as lambda *b2*, some of the l strands of lambda will anneal to the r strands of lambda *b2* and vice versa. The *b2* mutation constitutes a deletion of DNA from the lambda molecule (see Chap. 5). At the site of the deletion, there is some extra DNA in the *b2+* strand of the hybrid molecule which has no partner in the *b2* strand and therefore forms a single-stranded loop, visible in the electron microscope. From the location of the loop, the distance from the *b2* marker to the ends of the DNA molecule can be determined.

Hybridization can likewise be performed between lambda and variants such as lambda *dg* (Chap. 8) and lambda *imm434* (Chap. 9). These variants do not have deletions but rather substitutions of heterologous DNA at locations that have been mapped genetically. The hybrid molecules in this case have two unpaired single strands in the heterologous region.

The method can be refined by initial separation of the strands with poly IG. Mixtures can then be made in which all reannealed molecules will be hybrid (e.g., l strands of lambda and r strands of lambda *b2*).

Molecular cytogenetics provides the most direct and precise method for determining the physical location of lambda genes. The technique is diagrammed in Fig. 2-5.

If lambda DNA is partially denatured by gentle heating in a formaldehyde solution, local denaturation of certain regions of the molecule occurs without complete strand separation (Inman, 1966). The sites of local denaturation are observable in the electron microscope and can serve as additional cytogenetic markers.

Physical and genetic information on lambda is summarized in Fig. 2-3 (adapted from Szybalski *et al.*, 1966). Assignment of messenger RNA to a particular time and location on the map is from Skalka (1966); assignment to strands is based on Hogness *et al.* (1966) and Szybalski *et al.* (1966).

COLIPHAGE P1

In 1928, Bordet and Renaux discovered that the Lisbonne-Carrere strain of *E. coli* was lysogenic. Plating of the supernatant of this culture on a susceptible strain resulted in plaque formation. The strain was occasionally employed in various experiments during the following two decades, but only following the impetus of Lwoff's work was a careful, genetically oriented investigation undertaken of the phages carried by this strain.

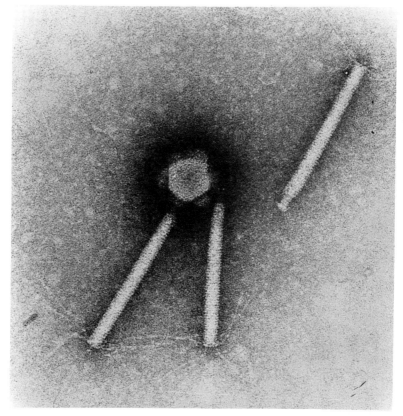

FIG. 2-6. Bacteriophage P1. Magnification 180,000×. Uranyl acetate negative staining. (Electron micrograph kindly supplied by F. Eiserling.)

In 1951, Bertani found that the Lisbonne strain was in fact lysogenic not just for one phage type but for three phages, all of which can form plaques on *Shigella dysenteriae*. These phages were called P1, P2, and P3 (P for phage). P3 has been involved only in very occasional studies since, but both P1 and P2 have been used extensively.

P1 has a particle weight of about 125×10^6 daltons, of which DNA comprises 58×10^6 daltons. The head has a diameter of 900 Å, and the tail measures 2200×200 Å.

Not much is known about P1 genetics. Little work has been done with plaque morphology mutants because the plaques on most indicator strains are small and poor. Isolation and mapping of conditional lethal mutants are underway at the time of writing (Scott, 1968). Phage P1 does not qualify as a proven episome because attachment to the bacterial chromosome has not been demonstrated. A piece of the bacterial chromosome (the *lac* region) can occasionally become attached to the phage genome.

The main reason P1 has been so widely used is that it mediates generalized transduction. Any phage lysate contains some abnormal phage particles which carry fragments of the bacterial genome, whose transfer to other bacterial cells is recognizable if the recipients are suitably marked genetically. Whereas lambda phage occasionally incorporates bacterial genes near the locus of prophage insertion, P1 can transfer genes from any part of the chromosome. The transduction of *lac* alluded to above has more in common with specialized transduction of the lambda type. Both phenomena will be treated in more detail in Chap. 8.

An interesting property of P1 phage is its ability to cause DNA modification and restriction. The presence in a cell of phage P1 somehow renders that cell incapable of accepting foreign DNA (either bacterial or viral) unless that DNA has been formed in the presence of P1 phage. Thus, lambda phage grown on an *E. coli* strain nonlysogenic for P1 forms plaques on a P1 lysogen only with very low efficiency. In a bacterial cross between a nonlysogenic donor and a P1-carrying recipient, both the amount of effective gene transfer and the degree of genetic linkage of donor markers are greatly reduced from normal (as though that DNA which did manage to function in recombination had been degraded into small pieces).

Ability to cause DNA modification is under the genetic control of the P1 phage. Mutants are known that do not discriminate against foreign DNA. Some of these are still able to "modify" DNA made in their presence, so that it is not rejected by normal P1 lysogens; others have lost both "rejecting" and "modifying" ability. Such mutants are still normal in their phage properties, including superinfection immunity.[3] DNA modification can also be caused by bacterial genes not known to be prophages.

---

[3] These mutants are picked as prophages and therefore are selected for retention of superinfection immunity. We can assert that mutations affecting DNA restriction but not immunity can occur. We cannot say whether these mutants are atypical or a very special class of all restrictionless mutants.

The stocks of phage P1 used in various laboratories are all descendants of the phage Bertani isolated from the Lisbonne strain. When Lennox (1955) employed this phage for transduction studies in *E. coli* K-12, he isolated a double mutant which (1) plated with high efficiency on all his K-12 stocks (*k*) and (2) formed clearer plaques (*c*) and gave higher titers than wild P1. The P1 *k c* stock has been used for transduction studies. My own experience has been that stocks labeled P1 *k c* are frequently heterogeneous mixtures of clear and turbid plaque-formers and have doubtless diverged considerably from Lennox's original isolate.

Among a group of phages isolated from hospital strains of *Escherichia coli,* one (called phage 363) was capable of generalized transduction, and has been employed extensively by the Pasteur group. It has the same lysogenic immunity as P1, and is serologically related to it.

COLIPHAGE P2

Of the three phages carried by the Lisbonne strain, Bertani selected P2 as the one most suitable for detailed studies about lysogeny. The phage forms plaques on various strains of *E. coli* as well as on *Shigella dysenteriae.* The free phage is similar to lambda in size and X-ray sensitivity. Although good plaque morphology markers have been available for many years, the rarity of vegetative recombination has precluded genetic mapping. Recently, numerous temperature-sensitive mutants have been isolated, and a map comparable in detail to that of lambda should soon be available (Lindahl, 1967; Bronson and Kelly, 1967).

As a temperate phage, P2 differs from λ in some important ways:

1. Whereas lysis of lambda lysogens can be induced by various agents, P2 prophage is noninducible. This difference need not have a complex basis, as lambda can mutate to noninducibility.

2. Lambda prophage occupies a unique site on the bacterial chromosome, between *gal* and *bio*. In polylysogens carrying more than one lambda prophage, the additional prophages are also located at, or very near, the same site. P2 prophage attaches to the chromosome at a particular site, near *his*. However, in double lysogens produced by superinfection, the second prophage is located at some other site, frequently far removed from the first one.

This second difference also need not be a fundamental one. Evidence is beginning to accumulate that, in special circumstances, P2 can form tandem double lysogens of the lambda type (Six, 1968). It is also possible, by some genetic chicanery, to create bacterial strains where lambda can add at more than one chromosomal site.

When P2 infects the B strain of *Escherichia coli,* some phage are produced that differ from P2 in several characters, one of which is their immunity specificity. These phage cannot form plaques on *E. coli* B, but will grow on *Shigella* or on P2 lysogens thereof. Apparently *E. coli* B harbors a genetic element capable of recombining

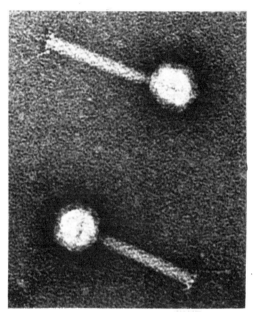

FIG. 2-7. Bacteriophage P2. Magnification about 200,000×. (Electron micrograph kindly supplied by Dr. T. F. Anderson.)

with an infecting P2 particle to produce a phage whose immunity specificity differs from that of P2, just as the immunity specificity of lambda differs from that of 434.

SALMONELLA PHAGE P22

In 1951, Zinder and Lederberg traced the production of bacterial recombinants between *Salmonella* strains to the agency of a virus harbored as prophage by one of the stocks used. This phage was called P22 or PLT22 (because it grew on the LT strain of *Salmonella*).

Phage P22, like coliphage P1, has served as a workhorse for bacterial geneticists not too concerned with the phage itself. However, from work of Zinder, Levine, Hartman, Thomas, and others, substantial information, both genetic and physical, is now available. The phage particle consists of (1) a head containing DNA of molecular weight 27×10^6 daltons and about 20×10^6 daltons of protein and (2) a short protein tail consisting of six spikes attached to a central core.

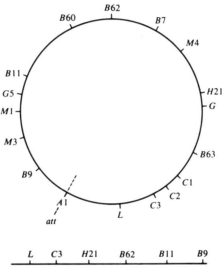

FIG. 2-8. Genetic map of phage P22 (above) and its prophage (below). Markers labeled B and G are heat-sensitive mutants. M and H are plaque morphology markers. $C1$ and $C2$ are genes analogous to the cII and cI genes of lambda. Gene L is necessary for prophage insertion, like the int gene of lambda (see Chap. 6). $A1$ determines ability to cause antigenic conversion. The att site (not precisely located with respect to $A1$) is the place at which the phage chromosome is broken during insertion into the bacterial chromosome (see Chap. 5). [Based on H. Smith (1968) and Young *et al.* (1964). Used by permission of the authors.]

The total weight of the six spikes is about 1.2×10^6 (Israel *et al.*, 1967).

The physical data indicate that P22 DNA molecules comprise a collection of linear molecules with circularly permuted nucleotide sequences and some terminal redundancy. The genetic map is circular (Fig. 2-8). Available data suggest that this phage, like lambda, is genetically differentiated into distinct functional regions. Analysis of conditional lethal mutants should make detailed assignments possible in the near future.

When phage P22 lysogenizes, it attaches to the *Salmonella* chromosome between the *proC* and *proA* markers. This location, like lambda's is unique, although special tricks can be used to force attachment elsewhere.

A cluster of three genes for clear plaque formation, analogous to those of lambda, have been characterized. Heteroimmune relatives of P22 were described by Zinder (1958). Interestingly, the immunity determinants are not allelic to any of the clear genes but rather are linked to the plaque morphology marker M3. Both mutants defective in insertion and natural variants with altered insertional specificity have been described (Smith and Levine, 1967; Young and Hartman, 1966).

Table 2-1 summarizes some properties of the four phages described here.

OTHER TEMPERATE PHAGES

We have described four representative temperate phages which have been used to elucidate some properties of episomes and which (except for P1) have been extensively characterized genetically. The number of known temperate phages is of course much larger, and not all of them need fit the same rules followed by the representatives chosen here. In particular, chromosomal attachment, while necessary for an element to be called an episome, need not be an attribute of all temperate phages.

Many interesting examples, some of them still incompletely studied, suggest that a permanent "phage carrier" state, without chromosomal attachment, may not be uncommon. The demonstration of such a state has generally been complicated by the possibility that virus liberated by some cells of the culture might be continually reinfecting others. Reinfection implies formation of mature extracellular particles, expected to be sensitive to antiphage serum.

Under certain conditions, some of the small single-stranded DNA and RNA viruses do not cause cellular lysis but appear to be secreted gradually from infected cells over a period of time. Indefinite perpetuation of this condition frequently is inhibited by antiserum, but in some cases seems not to be (Hsu, 1967). Serum sensitivity does not demonstrate an extracellular phase; but serum insensitivity, if rigorously documented, excludes it.

Some of the common *Bacillus* phages (in particular, phage SP10) appear also to be perpetuated in an unstable "carrier" state, where reinfection is necessary for permanent maintenance of the condition. The ability to produce phage can be transmitted through the spore

TABLE 2-1. Properties of Temperate Phages[a]

Phage	Common Hosts	Attachment to Bacterial Chromosome	DNA Molecule	
			Topology	Molecular Weight
Lambda (λ)	Escherichia coli	Extreme preference for one site; double lysogens usually carry both prophages at same site	Unique linear sequence; complementary single-stranded ends	33×10^6
P1	Escherichia coli Shigella dysenteriae	Undemonstrated	Population of molecules with circularly permuted sequence. Double-stranded terminal repetition[c]	58×10^6
P2	Escherichia coli Shigella dysenteriae Serratia marcescens	Strong preference for one site; double lysogens usually carry prophages at different sites	Unknown	24×10^6 [e]
P22	Salmonella typhimurium	Extreme preference for one site	Population of molecules with circularly permuted sequence; double-stranded terminal repetition	27×10^6

TABLE 2-1. (Continued)

Phage	Genetic map		Inducibility by Chemicals or UV	Related Heteroimmune Phages[b]	
	Autonomous Phage	Prophage		Same Attachment Site(s)	Different Attachment Site(s)
Lambda (λ)	Linear	Linear. (Cyclic permutation of phage map.)	+	434 82	21 80 424
P1	Probably linear[d]	Unknown	−		
P2	Linear[f]	Linear. (Cyclic permutation of phage map.)[f]	−	P2HyDis	
P22	Circular	Linear. (Linear derivative of phage map.)	−		P221

[a] References are included only for those data too recent to have been reviewed in the text.
[b] "Related" here means "able to recombine."
[c] Thomas (1968).
[d] Scott (1968).
[e] Mandel (1967).
[f] Calendar and Lindahl (1968).

stage and acquires the heat resistance typical of spores, which was one of the classical arguments for the existence of prophage as an intracellular, self-reproducing entity different in nature from the phage particle itself. But there is no evidence that this phage ever is integrated into the bacterial chromosome (Bott and Strauss, 1965).

As has happened in many areas of science, in the initial phases of the modern era of lysogeny certain very general questions were asked of quite specific material. First, Lwoff defined the prophage as an intracellular form in which phage specificity was reproduced, and demonstrated that such an entity existed. For technical reasons, this demonstration was performed on a lysogen which showed an exceptionally high rate of phage liberation, and whose prophage may very well not have been attached to the bacterial chromosome. Next, it was asked what was the nature of prophage. The answer, for technical reasons, was obtained with lambda and P2: Prophage is chromosomal.

The question of the ultimate stability of a nonchromosomal prophage is an interesting evolutionary one. On the practical side, there is no serious doubt that phage can under some circumstances be multiplied over at least a limited number of cell generations in cells that are growing and dividing. How stable such a system may be in the complete absence of intercellular phage transmission is uncertain.

In the next two chapters, we shall deal with bacterial transfer agents, some of which are episomes, and with nonepisomal plasmids. For these systems, indefinite perpetuation of the nonintegrated state seems to be the rule. The stability of transfer agents in the complete absence of reinfection (in this case by cellular contact rather than by liberation and reattachment of free particles) is not definitely established in all cases.

3

TRANSFER AGENTS

One motivation for the episome concept as originally propounded was to focus attention on certain apparent similarities between temperate phages, the fertility agent F, and the colicinogeny determinants. This chapter will treat those elements, including F, that cause bacterial conjugation. These elements can be called sex factors, fertility agents, or *transfer agents*. In general, conjugation results in frequent transfer of the agent itself and rarer transfer of other nonchromosomal elements, the bacterial chromosome, and portions thereof.

Only a few of these elements have the demonstrated ability to add to the bacterial chromosome (and therefore the right to be called episomes). We discuss them together because of their similar properties, just as we treated phage P1 with other temperate phages although its chromosomal attachment is not documented.

THE FERTILITY AGENT F OF *E. COLI* K-12

As mentioned in Chap. 1, direct information on the physical and genetic properties of F has accumulated less rapidly than that on phage because of the technical difficulty of separating F from the cell that harbors it. Nevertheless, we now have an approximate idea of the molecular weight of F DNA (35×10^6 daltons). Like the DNA of phage lambda, F DNA appears to be differentiated into discrete regions of different base composition. About 10 percent of the molecule is 44 percent guanine + cytosine (GC), and the other 90 percent is 50 percent GC. About 50 percent of F DNA is hybridizable with that of *E. coli* (Falkow *et al.*, 1967). These studies have been greatly expedited by the fact that F and other transfer agents can infect genera other than *Escherichia* that have a different average base composition. F DNA is separable from such host DNA by density gradient centrifugation.[1]

Genetic variants of F altered in ability to replicate or to initiate efficient transfer have been reported (Cuzin and Jacob, 1965; Hirota et al., 1966). Complementation tests show that several distinct functions are needed for manifestation of the normal F character. Genetic recombination presumably can occur, but no detailed map is yet available.

Operationally, the F agent is manifested by the ability of the cell containing it to function as a genetic donor. It also renders the cell sensitive to a group of small bacteriophages to which it is otherwise resistant. Both properties seem to depend on cellular appendages known as pili. These are about 85 Å in diameter and 1–20 microns long. Of the various types of pili formed by enteric bacteria, F-pili are distinguishable by the specific adsorption of male-specific bacteriophages to their surfaces. Aggregation of phage particles around F pili is easily visible in the electron microscope (Brinton, 1965).

In all probability, F pili are the true "conjugation tubes" through which DNA passes from donor to recipient in bacterial mating. The structure of the pilus is similar to that of a phage tail, with an axial hole 20–25 Å in diameter, well suited for DNA conduction. When cells are made to lose their pili, either by mechanical shearing or by aging the culture, donor ability disappears until conditions suitable for pilus regeneration are provided.

This makes F seem more and more like a phage. The main difference between a transfer agent like F and a temperate phage like lambda is from one viewpoint trivial: Lambda synthesizes a protein coat to enclose its DNA, attaches a tail to it, then lyses the cell and finds another host. F synthesizes only the "tail" and somehow implements its extrusion through the cell wall and membrane so that it converts the entire donor cell into an infectious particle suitable mainly for transferring F DNA to another cell.

If the pilus were to break away from the donor cell, with the F DNA inside, before contact is established with the recipient, it would thereby become a virus—and this would not appear to require any very complex change in properties. (Whether it does depends on the

---

[1] Transfer to other genera is not easily detectable for F itself, but only for derivatives of F (F') that carry some bacterial genes as well (see Chap. 8 for origin of F'). It is actually the F' DNA that has been separated. The principal technical problem in this area has been to distinguish that fraction of the F' DNA which is specific to F (Falkow and Citarella, 1965).

role of the donor cell in F-transfer, which we will discuss in Chap. 10.) The filamentous F-specific DNA phages, which seem to be "all tail" are intriguing objects for comparative studies.

Whereas production of a phage like lambda is lethal to the cell, the donor of F does not die, but continues to multiply indefinitely while harboring F. F therefore also multiplies indefinitely in the F^+ culture. The stability of the F^+ character may be partly due to intercellular reinfection of spontaneous F^- segregants. However, we shall see later that some transfer agents can give rise to mutants completely defective in transfer but still able to be maintained.

The ability of F to attach to the bacterial chromosome is thoroughly established, as we have seen in Chap. 1.

Operationally, a bacterial culture is termed F^+ if it will mate with an F^- tester strain. The fertility of *E. coli* K-12 and its substrains derives from the agent we call F. F^+ stocks used in different laboratories all harbor an agent descended from the original K-12 "F," although each may have evolved to some extent from that condition.

Such evolution can occur not only by simple mutations or deletions, but also by addition of blocks of genes from the bacterial chromosome. This phenomenon ("gene pickup") is known for most demonstrated episomes (see Chap. 8). Essentially, it results from errors in release of the integrated element from the chromosome, so that it takes with it some neighboring host genes. Variants of F deliberately selected for pickup of particular genes are in common use. Where the molecular weight of such variants has been measured, it has proven to be larger than that of standard F.

Bacteria other than *E. coli* K-12 have on occasion been screened for transmissible fertility. Many other stocks of the Enterobacteriaceae were tested by the Lederbergs for ability to mate with a K-12 F^- strain. Some of those strains that were fertile with K-12 F^- proved to harbor agents similar to (but not identical with) the F factor of K-12.

These other agents have also been referred to as F agents, and the cells that carry them, as F^+. This is a rational procedure, provided that the use of a common term does not prejudice the question of whether different "natural" F agents are related to each other, and to what extent. There is indeed some advantage (as we shall see presently) to reserving the term "F" for the standard element present in *E. coli* K-12.

COMPOSITE TRANSFER AGENTS

In Chap. 1, I listed those features of the bacterial mating system most pertinent to the postulation of a discrete genetic element or particle called the F agent. Conceptually, F was originally a logical abstraction: a purely formal entity, whose postulation introduced a convenient economy into the description of conjugal ability and its inheritance. The great power of such abstractions is witnessed by the success of genetics throughout its history.

In the years that have intervened since the discovery of F in 1953, identification of the abstract factors of bacterial geneticists with real physical structures has progressed remarkably. When we speak of a genetic element, we assume that we are talking about some nucleotide sequence. The assertion that an element is not attached to the chromosome implies that it constitutes a separate DNA molecule.

The idea that "autonomous F" is really a separate molecule in the cell still rests on indirect evidence. This has not created any serious problem in our discussion of F. So long as we are concerned with an element that is identified by a single property, and so long as we understand the operations by which it is defined, we can fruitfully operate with the abstraction at the same time efforts are in progress to discern its exact physical basis.

The identification of genetic elements with nucleic acid molecules has encouraged development of a coherent theory of fine structure genetics. This theory has the firm expectation that any structure qualifying as an autonomous genetic element should be sufficiently complex to be divisible into subunits that are mutationally and recombinationally distinct. The "unit factor" of the classical geneticist is replaceable by the muton, the recon, the cistron, or even a collection of linked cistrons, each in the appropriate operational context. We have used the term *F agent* rather than *F factor* in anticipation of a nomenclatural confusion that otherwise can develop when an agent is dissected into its component parts.

It thus came as no surprise when it was found that the F agent could mutate in a variety of ways, and that different mutants could show functional complementation with each other. This was exactly what was expected.

The isolation of mutants of F, however, raises a new question. So long as only wild-type F was available, its behavior could be followed only by its presence or absence in a given cell. Analysis at this level led to the postulation of an autonomous F agent. If fertility depends on two mutually independent determinants, we can verify that each in turn behaves as an autonomous element. This creates the formal genetic question of whether the two determinants are connected on the same genetic element or belong to different elements. The equivalent physical question is whether the two determinants constitute portions of the same or different nucleotide sequences.

A genetic test of this question must consist in some sort of linkage analysis. The two mutant agents must be introduced into the same cell. Then, in some manner, progeny must be examined to see whether the determinants remember their parental associations rather than assorting at random. Insofar as this type of analysis has been applied to the F agent, the results strongly indicate linkage. So components of transmissible fertility seem to reside on a single element.

Actually, with F itself this question has created no real problem. We expect a single element to consist of subunits. The finding of subunits does not impel us to worry about multiple elements, except as a formal possibility.

We shall deal next with a group of agents in which the situation is less straightforward. These are agents which determine, in addition to conjugal fertility, other properties such as antibiotic resistance or colicin production. It is pertinent to ask whether fertility determinants and resistance determinants are located on the same or different structures.

The facts about resistance agents will be detailed in the next section. We shall see that, by most criteria, resistance genes and fertility determinants behave as though linked on a common structure. On the other hand, certain groups of (individually mutable) determinants can be lost from the cell or transferred to another cell as a unit. Under these circumstances, some investigators elect to describe their findings in terms of separable, interacting units. Others treat essentially the same facts as though a single composite element were involved.

We shall find ourselves unable to resolve the situation completely. To avoid confusion, we prescribe in advance the following nomenclature: We will use the name R (resistance) agent to include all the drug resistance determinants and also those genes mediating conjugation and gene transfer. A cell harboring such an agent is an R^+ cell. Because an R^+ cell can lose some or all of the resistance determinants but still retain transfer ability, that part of the agent specifically concerned with transfer is called the resistance transfer factor (RTF), as distinguished from the complete R agent. The R agent as a whole is a (composite) *transfer agent,* which includes as parts the transfer factor and the resistance determinants.

RESISTANCE TRANSFER AGENTS

Among those agents that might have been isolated as "natural F agents" are some that were discovered originally because of their ability to confer drug resistance on cells carrying them. Their history and properties have many ramifications, epidemiological and ecological as well as genetic and physiological.

These agents were first noticed in Japan in the early 1950s. This was in the heyday of the wholesale indiscriminate administration of antibiotics for all ailments that might respond to them. Antibiotic therapy against dysentery was at first effective. According to good microbial genetic doctrine, it should be especially so when more than one unrelated antibiotic is administered simultaneously, so that no single-step mutant can survive. From about 1955 onward, however, there appeared an increasing number of cases in which the bacteria were simultaneously resistant to several different antibiotics for which genetic cross-resistance had not previously been observed—streptomycin, tetracycline, chloramphenicol, and sulfonamide. The first reported case was a patient who had just returned from Hong Kong; that seems to be as far as the origin has been traced (Watanabe, 1963).

At any rate, multiple drug-resistant shigellae became very common in Japan after that time. To explain their epidemiology, Akiba made the (at the time) bizarre suggestion that multiple drug resistance could be transferred from the intestinal *E. coli* of the patient to the infecting *Shigella* cells. It was soon demonstrated, both *in vivo* and *in vitro,* that this could indeed happen. The character of multiple drug resistance showed an infectious spread through the bacterial population, similar to that seen with F.

Because of their epidemiological importance, numerous natural isolates of R agents have been characterized, and some interesting differences have been noted among them. Intensive study has centered around a relatively small number of isolates. We shall return to the natural variability and epidemiology of R factors presently. First, a more detailed description of the properties of particular isolates is in order.

THE R AGENT OF *SHIGELLA FLEXNERI* 2b STRAIN 222

This R agent, isolated by Nakaya in 1960, has been the object of detailed study since around 1960 by Watanabe and his colleagues, and their stocks have been used by workers in various other laboratories. It contains genetically separable determinants for resistance to four drugs—streptomycin (*str*), tetracycline (*tet*), sulfonamide (*sul*), and chloramphenicol (*cam*). It also causes resistance to mercury ion and to some aminoglycoside antibiotics not yet in current use (D. Smith, 1967b).

Cells harboring R can transfer it to other cells. They also transfer chromosomal genes, but at a much lower frequency. This shows that they carry a transfer factor, associated or interacting with R. They form pili very similar to F pili. If these cells are to be classified as F^+ or F^- on the criteria originally used, they must be considered F^+. They do not harbor the K-12 F agent as such, but an element that interacts with it. When the R agent is transferred into an F^+ cell, conjugal fertility drops from the normal F^+ level.[2] This interaction seems to concern specific repressors of pilus formation, and will be discussed in more detail in Chap. 9. The interaction is phenotypic, not genetic. If the cell carrying both elements loses R, it behaves again as a normal F^+.

The factor responsible for transfer (as opposed to the whole complex including the resistance agents) was called the resistance transfer factor (RTF). It has also been treated as the mating gene (*m*). Operationally, the two terms are identical. Presence of RTF (m^+) denotes transfer ability. Absence of RTF (m^-) denotes its absence or lack of function. Ability to transfer resistance determinants is thus far genetically inseparable from ability to transfer the bacterial chromosome (Hirota et al., 1966).

In principle, the transfer factor RTF and the resistance determinants for individual drugs might be located on the same genetic (and, by implication, physical) structure, or they could be separate but interacting units. To distinguish these alternatives, Watanabe and Fukasawa (1961) first transferred the R agent into *Salmonella typhimurium* and *Escherichia coli* and then transduced drug resistance from these donors to sensitive recipients, using phage P1 in *E. coli* and phage P22 in *Salmonella*. The idea was that linked transduction of different properties should mean that they depend on different parts of a common structure.

The result was different in the two phage-host systems. With P1 on *Escherichia coli*, over 80 percent of transductants selected for resistance to any one antibiotic were resistant to all the others. Among those selected for tetracycline resistance, about 10 percent were re-

[2] For technical reasons, a drop to zero fertility is not readily distinguishable in this system from a drop to the level characteristic of R^+ bacteria.

sistant to only tetracycline and 2 percent to the three-drug combination streptomycin-chloramphenicol-tetracycline. All these transductants were again able to transfer drug resistance into suitable recipients.

In *Salmonella,* phage P22 (which contains less DNA than phage P1 and therefore is expected to transduce smaller pieces of DNA) never transduced the whole four-gene complex. The gene blocks transferred were *sul-str-cam, sul-str,* and *tet.* Of 184 transductants observed, only one of the 49 *tet* transductants was able to transfer drug resistance.

Because of the cotransduction of all properties by phage P1, it was inferred that the whole unit, including the transfer factor, constituted a block of linked genes. The patterns of fragmentary transfer with the two phages were both compatible with the order

sul - str - cam - (RTF, *tet*)[3]

This map was expanded by Hashimoto and Hirota (1966). They isolated several nonallelic mutant R agents that were unable to confer *cam* resistance, and corresponding mutants for tetracycline and sulfonamide and streptomycin. (Resistance to the last two could be lost or regained by a single mutation.) Simultaneous infection with two genetically marked R agents produced an unstable diploid condition, with eventual segregation of parental and recombinant types. The *cam* determinants mapped in a cluster linked to *sul-str. Tet* was linked to transfer ability, but there was no clear linkage between these determinants and the rest of the map.

Numerous independent transfer-deficient mutants have been isolated. Some cooperate with each other physiologically, but no genetic map is available. Introduction of the F agent into such strains (either free or chromosome-bound as an Hfr) restores transfer to its normal R frequency in some (but not all) cases. Likewise, in Hfr bacteria deficient in mating ability, introduction of an R agent restores virility. When a nontransmitting R is transferred from an Hfr donor, it remains nontransmitting in the new host, indicating that F is supplying some physiological transfer function rather than physically attaching to R.

Spontaneous segregants lacking the complete resistance of the parental strain can be found. The four determinants are commonly lost as a unit. Partial losses also occur. Either the *tet* gene alone or the block *sul-str-cam* can be deleted. Other combinations are rare. Drug-sensitive segregants seem to have suffered physical loss of some R-specific DNA (Falkow *et al.,* 1966). Segregants resistant to three or

[3] The order of RTF and *tet* with respect to the other markers cannot be determined from the data.

four antibiotics but unable to transfer resistance are also encountered (Watanabe, 1963). When an R agent bearing the *tet* determinant was introduced into these cells, all resistances became again transferable, and stable "recombinant" R agents carrying all four resistance factors in their normal configuration (as judged by transduction) were reconstituted. These segregants can be interpreted as resulting either from loss of RTF with retention of the resistance determinants, or from a mutation of RTF that destroys transfer ability. The only rigorous conclusion is that transfer ability is not required for replication of resistance determinants.

No good evidence for chromosomal attachment of the R agent has been obtained. It cannot properly be called an episome as yet. Chromosomal attachment of RTF without the drug determinants has not been examined. Attachment of drug determinants without RTF does occur in P22 transduction. As mentioned earlier, Watanabe and Fukasawa (1961) found that P22 transduction produced individuals carrying only some of the drug resistance determinants, and unable to transfer these by conjugation. Their drug resistance is also not curable by acridines (Harada *et al.,* 1963). Unlike the cases treated in the previous paragraph, in these strains drug resistance did not become transferable on infection with a normal transfer factor. The determinants were therefore postulated to have become incorporated into the bacterial chromosome, perhaps by replacing homologous genes there.

Dubnau and Stocker (1964) showed that resistance determinants for tetracycline can become integrated at the attachment site of phage P22, in a recipient of P22 transduction. The resistance determinants appear in this case to have been incorporated into a transducing phage.

Strict adherence to definitions requires us to say that, though R is not an episome, the resistance genes are. Any other course would presuppose more knowledge than is actually available, and would imply that the role of the known element P22 in attaching R genes is not played by some unknown element in cases where we *think* that the element attaches "all by itself." The difficulty is that the definition of the episome we are using is operational rather than biological. We shall return to this point in Chap. 13. Whereas chromosomal attachment has not been demonstrated, Pearce and Meynell (1968) showed that an R agent could mobilize the bacterial chromosome from a specific point of origin. Possible mechanisms of chromosome mobilization will be discussed in Chap. 4.

I think it is unlikely that the resistance determinants have any chromosomal homologues in the hosts where they are now found. In the first place, the physiology of the resistance characters is generally different from that displayed by known chromosomal mutants. Resistant mutants of the host are frequently recessive. Resistance determined by an R agent must be dominant to or epistatic over the chromosomal genes causing sensitivity to the same antibiotic; otherwise it would never be observed. Dominance of *cam* resistance to *cam* sensitivity has been shown in the R agents themselves. There are now several cases in which convincing or circumstantial evidence indicates

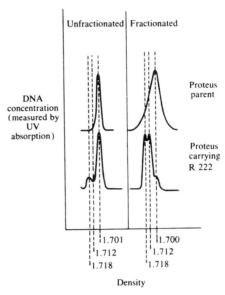

FIG. 3-1. Detection of R factor DNA in *Proteus mirabilis*. The upper left curve shows the UV absorption of *Proteus mirabilis* DNA, with a single mode at a density of 1.701. Below is the curve for the same strain after conversion to drug resistance by R infection, showing a prominent satellite band at density 1.718. The right curves show the result of rerunning the denser fractions of both DNA samples to enrich for denser DNA. Two satellite bands are now visible for the culture carrying R, at 1.718 and at 1.712. When DNA of the parent culture is similarly fractionated, no satellite bands appear (upper right). [Taken from Rownd *et al.* (1966). Used by permission of the authors.]

that R agents cause resistance through synthesis of enzymes that inactivate the antibiotic. This is best documented for penicillin (Datta and Richmond, 1966) and chloramphenicol (Shaw, 1967). Furthermore, molecular studies (to be discussed later) suggest that the DNA comprising the resistance determinants does not have the same base ratios as host DNA. All available evidence indicates that R agents are basically extrinsic, extrachromosomal elements.

The presence of R in a culture is accompanied by the physical presence of DNA with a distinctive GC content (Fig. 3-1). From density profiles of R-infested cells of *Escherichia, Proteus,* and *Serratia* (whose chromosomal DNA's differ widely from one genus to the next) it can be seen that presence of R is accompanied by two new bands. About 85–90 percent of the satellite DNA is 56–58 percent GC, and the other 10–15 percent is 50–52 percent GC. Assuming one copy of R per bacterial chromosome, in *Escherichia coli,* a weight of 25×10^6 daltons is calculated.

Because the DNA is extracted in such a manner that shear degradation reduces the average molecular weight far below 25×10^6, the two satellite bands could plausibly represent different fragments of a single molecular species.

If there is one copy of R per bacterial chromosome in *E. coli* and *S. marcescens,* the size of the satellite band in *P. mirabilis* indicates the presence of about 10 copies per chromosome there. A distinct satellite band such as seen in Fig. 3-1 is not expected in unfractionated samples if the amount of satellite DNA constitutes less than about 5 percent of the amount of host DNA. This provides an important technical limitation on this type of work, and also underscores the fact that the absence of satellite bands constitutes no evidence for the absence of undetected plasmids or episomes.

Segregants that have lost resistance genes can also show alteration of band patterns. Loss of the whole 56–58 percent band can accompany loss of the block *sul-str-cam* (Falkow et al., 1966); but such loss has also been observed in genetically complete R's, suggesting that this (major) fraction of the satellite DNA contains neither the resistance genes nor transfer activity (Rownd et al., 1966).

The transduction studies of Watanabe and Fukasawa suggest that R has a molecular weight greater than P22 DNA (26×10^6) and less than P1 DNA (60×10^6). The lower figure is well within the limits of error of the estimated 25×10^6 daltons. However, if all the pertinent genes are on the 52 percent GC piece, which should have a molecular weight less than 4×10^6 daltons, the discrepancy is quite large.

The R agent described in this section, coming from *Shigella flexneri* 222, has usually been designated "R222," but various workers have their own names for descendants of the same stock—such as R100, or NR1. The agent has been transferred through various hosts, and considerable variation may have accumulated. Under these circumstances, it is perhaps surprising that stocks of different workers behave as similarly as they do.

THE Δ FACTOR

In 1965, Anderson and Lewis investigated the effect of repeated passage of R agents from *Salmonella* to *Escherichia coli* and back. By this time, the incidence of drug resistance in British *Salmonella* isolates had risen to 61 percent (largely due to the presence of transmissible R agents), and one of these isolates (resistant to ampicillin, streptomycin, sulfonamide, and tetracycline) was used as starting material. The R agent in their strain differs from R222 at least by the fact that its presence does not interfere with expression of an F agent in the same cell. Of various R agents, some, like R222, show fertility inhibition (fi^+); others, like Anderson's strain, do not (fi^-). A property thus far inseparable from the fi character itself is the ability to cause DNA modification (characteristic of fi^- but not fi^+ factors, Watanabe et al., 1966).

Anderson's experiments had been preceded by an extensive survey of the types of drug-resistant strains in selected areas of Great Britain over a period of years. It was clear from these studies that different drug resistance characters had appeared successively in the population—*str* and *sul* resistance first, *tet* resistance later, and *amp* resistance later still. This makes it very likely that the individual resistance determinants in the stock studied had originally occurred singly rather than as a unit.

At any rate, Anderson and Lewis studied the behavior of the complex, once formed, on subsequent passage. Their results indicate what at first sight seems the exact opposite of Watanabe and Fukasawa's findings with R222. Instead of being passed from host to host as a unit, *amp* and *str* resistance were generally transferred individually, although there was some correlation between the two.

Transfer of tetracycline resistance, initially very rare, became so common in later cycles that essentially all the cells receiving *amp* or *str* received *tet* as well. This was observed in experiments where the period of contact between donor and recipient cultures was long (several hours). With shorter periods of exposure, cells could be isolated with only *amp* or *str*. These proved incapable of transferring resistance. In other words, there was independent transfer, not only of the different resistance determinants, but of the transfer factor as well. The latter finding was subsequently confirmed by isolation of strains harboring only the transfer factor and not the drug resistance determinants.

Anderson called the transfer factor Δ. This serves to emphasize its independence from the resistance genes, and also to distinguish it from the fi^+ RTF component of R222.

Cells that have received only the *amp* gene and not Δ can be made to transfer *amp* again by infecting them with Δ. The presence of Δ (independently of resistance determinants) can be diagnosed by this method. Δ is common among natural, drug-sensitive isolates. Δ also confers on the cell a difference in reaction to some of the *Salmonella* typing phages (perhaps related to the DNA modification caused by fi⁻ R agents) and can be recognized on this criterion.

This R agent is thus dissociable into several components—the individual resistance determinants and the transfer factor. The fact that a resistance determinant in a strain without Δ is readily mobilizable when Δ is introduced by infection suggested that this determinant can replicate autonomously. Similarly, since the determinant for *tet* was not ready mobilizable, it seemed that it might have originally been chromosomal and later have become tightly bound to Δ. Δ is then visualized as being able to attach to various resistance determinants, sometimes to more than one at once, and to conduct them across the conjugation bridge.

Anderson's R agent is of course an independent isolate from nature, and his results need not apply directly to R222. Rather than imagining entirely different mechanisms for the two systems (and rather than contemplating extreme possibilities or trivial explanations for either finding), we prefer to believe that both results mean what they seem and that different elements, initially separate, unite to form a common physical structure carrying both resistance determinants and transfer function.

The nature of the "attachment" between the drug determinants and Δ is left quite unspecified. Indeed, the facts do not require physical attachment between the two at all. If the drug determinants have a special propensity to pass through the conjugation bridges made by Δ, Anderson's observations would be explained. The physical data with R222 are not very illuminating with respect to this question, either. The solid evidence for attachment comes from the transduction experiments, and these, so far as they go, are explainable on a simple linear map. Both in Anderson's system and in R222, the linkage of *tet* to transfer function is much stronger than that of any other resistance determinant.

In the original statement of the episome concept, Jacob and Wollman did not specify the nature of the attachment, either. The important point was that two separate elements, episome and chromosome, could unite to produce a single structure. If Anderson's picture is correct, we would assume that attachment of Δ to R plasmids[4] follows the same basic mechanisms as attachment of F or lambda to the

---
[4] See Chap. 4 for definition and discussion of plasmids.

A

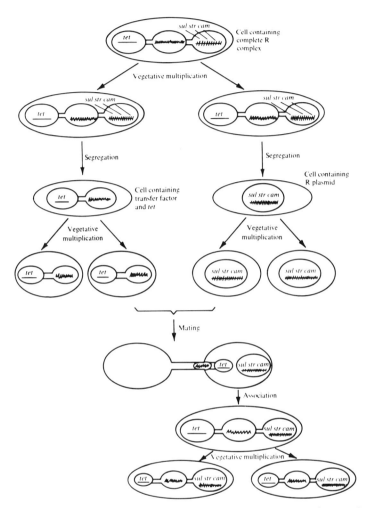

FIG. 3-2. Illustration of different descriptions of segregation and reassociation of characters determined by R elements. Both A and B represent the same sequence of observable events, visualized differently. In both cases, three distinct nucleotide sequences participate. *tet* is a sequence containing the gene for tetracycline resistance. ⁓⁓⁓⁓ is a transfer factor such as Δ or RTF. ⁓⁓⁓⁓ is a sequence containing genes for *sul, str,* and *cam* resistance.

In part A, these are depicted as three separate elements, each potentially autonomous, which are capable of association with one another in an unknown manner (depicted in the figure by surrounding each element with a circle and connecting these circles with double lines). At the outset we show a cell containing the associated elements. (The cell's chromosome is not drawn.) This complex divides as such, and each daughter cell contains the full complex. Occasionally, one or the other component fails to multiply (segregation), resulting in a cell harboring one or two elements of the complex rather than all three. Two independent segrega-

B

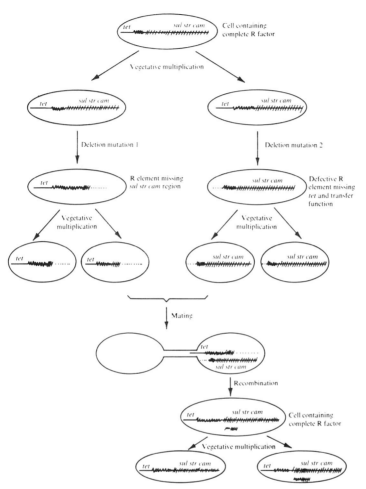

gational events are depicted in two different cell lines. As the individual elements are potentially autonomous, they continue to multiply as such in the two cell lines. If individuals of the two lines mate, the transfer factor and its associated element can enter the cell bearing the R plasmid, and reassociation restores the original element.

In part B, the three elements are arranged on a single linear structure which multiplies as such. Deletion mutations in different cell lines produce individuals that retain in one case *tet* and transfer ability; in the other case, *sul, str,* and *cam* resistance. The deletions are so situated that the two structures retain some genetic material in common. Mating introduces the two structures into the same cell, and allows recombination in the common region. The original structure is thereby reconstituted.

It is left to the reader to interpret the nature of the association depicted in part A in terms of known mechanisms to be detailed in Chaps. 5 and 6; and also to consider the location of the replication origin in both part A and part B (cf. Chap. 10).

bacterial chromosome. Δ would then be the episome of a plasmid. Our discussion of R agents and Δ in a book on episomes thus has more justification than mere similarity to the known episome F. Indeed, the question of whether Δ itself might attach to the chromosome has hardly been studied.

Many of the results with R222 fit well with Anderson's findings. The rare segregation of particular blocks of determinants (apparently accompanied by physical loss) is more easily described in terms of a separable determinant than a deletion mutation (although operationally there is little distinction). The rate of segregation depends very much on the host, *cam* resistance can barely be maintained in *Salmonella,* for instance. Anderson has described the R agent as an organized complex of dissociable elements. Formally, his results can equally well be ascribed to the existence of "defective R particles." Thus a cell that harbors the ampicillin resistance gene and cannot transfer it is defective in tetracycline resistance and transfer, whereas the cell with transferable *tet* resistance harbors a factor defective in *amp*. The fact that these two elements together can reconstitute the original structure can be described either as reassociation of independent elements (presumably by a recombinational mechanism, as discussed in Chap. 5) or as recombination between defective elements, leading to reassociation of the "nondefective" alleles. These possibilities are diagrammed in Fig. 3-2.

Clearly, much of the difference between the two descriptions is purely semantic. The most important operational distinction is whether the complex possesses more than one independent replication system. If the complex is formed by association of independent self-replicating elements (with no concomitant loss of material), then it must perpetuate the structural bases of the individual replication systems, although some may be repressed in the complex. At the moment, the facts serve mainly to illustrate the extreme difficulty of studying elements like F or R which are in no phase of their life cycle physically separate from the host cell.

It is not implicit in Anderson's concept that each resistance gene should necessarily be separable from all others. We would imagine that a block of closely linked genes such as *sul-str-cam* and R222 would be on one such R plasmid, and the *tet* gene on another.

Anderson's results have been interpreted in terms of evolution and origin of R agents as complexes of different, basically nonchromosomal elements. I find it hard to form any clear picture of the order of events over a long enough time span to be meaningful. What these findings mean to me is that association of and dissociation into component parts are features of present-day R agents as we find them—not of their distant progenitors in some shadowy past.

If this is so, we should consider the possibility that, as in the case of temperate phage attaching to the bacterial chromosome, association is something that happens at a definite time in the life cycle (cf. Chap. 6). Perhaps it is induced by conjugation, or by P1 infection.

EPIDEMIOLOGY OF R AGENTS

Because of their importance with respect to public health and chemotherapy, the levels of drug-resistant and R-carrying bacteria (especially *Salmonella*) have been monitored at various times and places, and many independent isolates have been examined. Factors are reported carrying resistance to various combinations of drugs in use (sulfonamide, penicillin, streptomycin, tetracycline, chloramphenicol, ampicillin, kanamycin, neomycin, polymyxin, radiomycin), drugs not yet in use (bluensomycin, spectinomycin, viomycin, gentamycin), metallic ions (mercury, cobalt, and nickel) and ultraviolet light (D. Smith, 1967$a,b,$). Except for the combinations viomycin-gentamycin and bluensomycin-streptomycin, resistance to each antagonist seems to be mediated by a separate gene.

Many of the natural isolates are doubtless closely related. Some of those carrying one or two resistance determinants behave identically to laboratory segregants from multiply resistant stocks. The two types fi^+ and fi^- seem to differ fundamentally in their transfer properties, their interactions with each other, and (by definition) with F. Thus far, no major variations in these properties have been found among different isolates of the same fi type. Recombination between F and fi^+ has been reported, but not between fi^+ and fi^-. However, the similarity in behavior of the individual determinants in the two types suggests that such recombination occurs in nature. Because of the complex nature of R agents, it is uncertain to what extent the recombinants reported should be regarded as associations of the same R plasmid with different transfer factors, rather than as conventional recombinants. Indeed, if the association itself has a recombinational mechanism, these two situations do not constitute distinct alternatives (cf. Chap. 5). Thus far, no fi^- agents have been reported to have the *cam* determinant or the 56–58 percent GC component.

The widespread incidence of R agents, at first detected in Japan, has now been recognized in many other parts of the world. Regardless of the ultimate origin of R agents, it seems likely that selection of the particular multiple resistance combinations found in a given area is largely determined by the presence of the relevant antibiotics in the environment of the bacteria in question. This situation is largely manmade. Antibiotics have been widely disseminated to human patients and have been incorporated into the diets of livestock as "preventive medicine" against possible drug-sensitive pathogens. The actual result is selection for a drug-resistant intestinal flora, and one which, unfortunately, is able to transfer this character to an entering pathogen—thus nullifying the efficacy of the antibiotic when the infection really develops.

Just as the epidemiology of human virus infections can depend on a reservoir of virus in animal populations, so the nonpathogenic drug-resistant Enterobacteriaceae can act as a reservoir for the R agents. From the basically "extrinsic" nature of the agents, it seems unlikely that the known enteric bacteria comprise the full extent of the reservoir. This population may in turn be exposed to rare infection with the real "natural" hosts of these determinants. But there is little doubt that mass dissemination of antibiotics has expanded the reservoir and made it potentially more dangerous to mankind.

Watanabe (1963) has given statistics on the rise of drug-resistant shigellae in Japan from about 0.2 percent in 1953 to 13 percent in 1960. In 1965, the level had risen to 58 percent (Mitsuhashi *et al.*, 1967). The large majority of these were resistant to the four drugs sulfonamide, streptomycin, chloramphenicol, and tetracycline and were able to transfer resistance to test strains in the laboratory. Eighty-four percent of the *Escherichia coli,* 93 percent of the *Klebsiella,* and 90 percent of the *Proteus* cultures were likewise resistant, and again the large majority of resistant individuals were carrying R agents. These statistics come from hospital patients and certainly exaggerate the situation in the population at large, but they are alarming enough in any event.

The increase in other countries has been no less striking. The common occurrence of R agents among the Enterobacteriaceae is well documented in Great Britain, Israel, the United States, Germany, and Switzerland (Datta, 1965). As in Japan, the frequency of strains carrying R agents has steadily climbed wherever measurements have been made.

There is no reason to suspect that R agents originated in Japan and spread from thence around the world. Studies of resistant strains isolated in the United States over a period of years indicate that most of them owe their resistance to R agents—and that this was true even before the chemotherapeutic use of antibiotics (D. Smith, 1966). Mutation and selection experiments involving pure cultures turn out to be rather poor models for predicting the response to antibiotics *in vivo*. The proper *in vivo* model is an epidemiological one in which fresh mutations play at most a minor and indirect role.

This might reflect some special features of drug resistance. On the other hand, perhaps we are seeing here the first illustration of the main natural mode of evolution of bacteria in response to major alterations in the selective properties of their environment. In no other case has the environment been subjected to deliberate alteration on such a grand scale with subsequent monitoring of natural populations to determine the extent of change and its genetic basis.

COLICINOGENY AGENTS

We have seen in previous sections that an R agent can be regarded as the combination of a cytoplasmic genetic element (R plasmid) with a transfer factor. The history of colicinogeny determinants has run a somewhat different course, but the same general picture seems to obtain.

Like R agents, colicinogeny determinants have been isolated from different natural strains of the Enterobacteriaceae, and have been introduced into common hosts for comparison. We again meet the same problem as with R agents: Where a determinant seems to manifest more than one property, how are we sure there are not really two separate elements simultaneously present?

R agents are obtained from drug-resistant isolates of common bacteria. Strains bearing R agents can be screened from those with chromosomal drug-resistant mutations because only the former are easily transmissible. Most multiple drug resistance characters prove to be transmissible, but naturally occurring nontransmissible R agents would generally have been screened out by this procedure. Presumably there should also be a strong selection in favor of infectivity for R agents from pathogenic bacteria of hospital origin.

With colicins, on the contrary, large numbers of strains producing different types of colicins had been collected and classified long before genetic studies were even contemplated. The material on hand should therefore constitute a fairly random sample of all those genetic determinants able to synthesize any product that is not lethal to the cells that elaborate it but specifically toxic to closely related bacteria. A survey of this material reveals these general facts:

1. There are many specific types of colicins. These are distinguishable by host range. As in the case of phage resistance, mutation to colicin resistance seems usually to alter the cell surface in such a

manner that the colicin is no longer bound. Cross resistances are found between particular phages and particular colicins, suggesting that agents of the two types can have common receptors. The attachment specificity of colicins is the basis of the standard classification scheme. Colicin E will attack one group of bacterial strains and colicin I, another. Mutants resistant to colicin E remain sensitive to colicin I and vice versa, indicating that the two colicins attach to different receptors.

As in the case of lysogeny, colicinogeny confers on the cell specific immunity to the colicin produced—even when that colicin can attach to the cell surface. Colicins with the same attachment specificities but different immunity specificities are distinguished as colicin E1 and E2 or colicin Ia and Ib.

To distinguish colicins with identical host range and immunity properties but of independent origin, the complete name of the colicin (and of its genetic determinant) includes the name of the bacterial strain from which it was originally derived. For example, colicin V-K94 is a colicin of type V made by strain K94, or by any other strain that has acquired its colicinogeny factor from strain K94 (Nomura, 1967).

2. Chemically, all known colicins are proteins, although the number of analyses of highly purified material are small. Colicins may sometimes be associated with nonprotein material, but the specific, biologically important component seems always to be protein. Some colicins have molecular weights as low as 60,000 and are thus typical small protein molecules, physically speaking.

There is another completely different type of structure produced by some bacterial strains that satisfies the formal definition of "colicin." These strains produce, either spontaneously or following induction with ultraviolet, particles with molecular weights of 10^7 or more which appear electron microscopically to be very similar to bacteriophage particles or to the DNA-free "ghosts" thereof. This type of colicinogeny is obviously a form of lysogeny for a phage that is defective either on the strain of origin or on the strains sensitive to it. Its inclusion with other types of colicinogeny is a semantic accident.

3. Genetically, all known colicinogeny determinants behave in an aberrant manner suggestive of extranuclear inheritance. Some determinants, such as those for colicins E1 and E2, behave like the nontransferring R plasmids. These factors are not by themselves infectious, but are transferable rapidly and independently in the presence of known transfer factors (either F or other colicin determinants). In other cases, the colicinogeny factor is infectious without added F.

Colicin V-K94. This determinant (Kahn and Helinski, 1965; Nagel de Zwaig, 1966) controls production of two separate colicins:

TABLE 3-1. Properties of Transfer Agents

Agent	DNA Molecule		Chromosomal Attachment	Curing by		Nature	Genetic Components	Interactions with F Agent		
	Molecular Weight	Topology		Acridines	Thymine Starvation			Interference with F Establishment	Repression of F Function	Complementation with Defective F
F	35×10^6	Circular[a]	+	+	+	Simple		+	−	+
R222	25×10^6	Unknown	−	+	+	Composite(?)	RTF $\Delta \cdot tet$[b]	−	+	+
$\Delta \cdot R$[b]	Unknown	Unknown	−			Composite	Δ R plasmid	−	−	Unknown
Col	Unknown	Unknown	Unknown (probably+)	+	+	Composite(?)	F	+	−	Unknown (probably+)
V-K94										

[a] Freifelder (1968).
[b] We use the symbols $\Delta \cdot R$ and $\Delta \cdot tet$ to refer to the R agent of Anderson and Lewis (1965) described in the text, and its derivative which tranfers only *tet*, respectively.

V and I. It has many similarities to the F agent, including synthesis of surface receptors (presumably pili) for male-specific phages, promotion of chromosomal transfer, ability to interfere with the establishment of the standard F agent. It is even possible to obtain from a col V agent mutilated by ^{32}P decay an apparently normal F agent that has lost all attributes of colicinogeny.

The reverse class (still colicinogenic, but no longer F^+) was selected for resistance to male-specific phages. These cells harbor a colicinogeny determinant that no longer imparts fertility, yet excludes standard F just as one F element interferes with another. This suggests that the F replication system is still being used and that a complete dissociation of a colicinogeny plasmid from transfer factor has not been achieved.

The col V complex, considered as a whole, is a transfer agent which contains F. F is capable of chromosomal attachment. Chromosomal attachment of the complex remains to be demonstrated but seems highly probable, as col V can become bound to chromosome fragments (Fredericq, 1963).

Colicin B. A col B agent, like col V, controls a transfer system apparently identical to that of the standard F agent. This agent can become integrated into the bacterial chromosome, forming an Hfr strain. The col B determinant and the F agent are jointly curable by acridines and are best considered as one unit (Fredericq, 1963). The unit is properly considered an episome.

Colicin Ib-P9. This factor resembles col V and col B in being autotransferable and able to promote chromosome transfer. Unlike the others, it has no specific similarity to F. It neither causes sensitivity to male-specific phages nor interferes with F if present in the same cell. F does influence col I transfer to some extent, however; and this col I agent can cooperate functionally with a defective F agent to restore normal transfer function. It presumably represents a different type of transfer agent, just as the fi$^-$ R agents do. It is not curable by acridines, but slight curing by thymine starvation is observed (Clowes *et al.*, 1965).

Nontransferring Colicinogens. The decomposition of colicinogeny transfer agents into transfer factor and independent plasmid has not been completely achieved. Other colicinogeny agents are independent plasmids, devoid of transfer ability. They are not known to attach stably either to the bacterial chromosome or to transfer agents such as F. These agents will not be described in detail.

The properties of some of the transfer agents described in this chapter are summarized in Table 3-1.

4

BACTERIAL PLASMIDS AND PARTIAL DIPLOIDY

Most known bacterial genes multiply as part of a single linkage structure, the bacterial "chromosome." Some groups of genes seem to multiply independently of it. These have been termed "plasmids." The special case of plasmids that can attach to the chromosome is the subject of this book. Those plasmids such as R222 and col Ib-P9 that form conjugation bridges, and those such as P1 that form heads and tails, have been treated in previous chapters.

Nonchromosomal elements are well known in many organisms. It would carry us too far afield to discuss the whole topic of cytoplasmic inheritance here. However, the situation in bacteria merits special attention. We have anticipated this in the previous chapter. Instead of describing particular plasmids in detail, we shall discuss here the operational criteria for determining "independent multiplication." These criteria are in general indirect and do not permit such rigorous conclusions as we might like.

ABERRANT INHERITANCE IN CROSSES

If we look at the assortment of genetic characters among the progeny of crosses, the chromosomal genes can be mapped on a single linkage structure. Plasmid genes, by definition, do not show linkage with any special part of the chromosome. This shows that the plasmid is transferred and integrated separately from the chromosome.[1]

[1] An alternative explanation for the absence of linkage would be that the plasmid is always transferred in association with the chromosome, but that it is connected to different chromosomal regions in different cells of the same culture. This alternative seems excluded for elements such as autonomous F whose transfer is much more rapid and frequent than that of any or all chromosomal genes.

The locus of multiplication, which we have included as part of our definition, is not clearly defined.

We have seen in Chap. 3 that some resistance agents, themselves devoid of transfer activity, are transmitted in crosses more readily than is the host chromosome. The same is true of many colicinogeny agents. Col E1-K30 and col E2-P9, for example, do not by themselves cause transfer, but are transferred at high frequency from donor cells in crosses, whether the donor ability is due to F or to another colicinogeny determinant such as col Ib-P9.

In these cases, we say that the transfer agent "mobilizes" the transferred element. Mobilization refers to the mediation by a transfer factor of the transfer of a second element to which the transfer factor is not directly attached at the start of the experiment. All known transfer agents can mobilize, at some rate, every other genetic element in the cell, including the chromosome. It is reasonable to expect independent mobilization of plasmids and chromosomes. Why plasmids should generally be mobilized more efficiently than chromosomes is not understood.

The transfer of chromosomal genes by wild-type F probably follows at least three mechanisms that differ at least in detail:

1. Because F can reproduce in the integrated state, any F^+ donor population will contain some clones of Hfr donors. Jacob and Wollman found that, under some circumstances, these account for a major fraction of the total donor ability of an F^+ culture.

2. The F factor may recombine with the chromosome, to create a structure equivalent to an Hfr chromosome, immediately prior to transfer. Much of the chromosome mobilization by F' agents (which have picked up some chromosomal genes in addition to the usual genes of F, see Chap. 8) is attributable to this source (Adelberg and Pittard, 1965). Mechanism (2) is distinguishable from mechanism (1) only if there is a specific enhancement in rate of recombination between the chromosome and the transfer agent at the time of mating. The recombination might depend on the bacterial "recombinase" system. Such a requirement has been reported (Clowes and Moody, 1966) and contested (Curtiss and Renshaw, 1968).

3. The chromosome may pass through the conjugation bridge made by F without any physical association between the two. This is suggested by the existence of F^+ strains that are unable to form Hfr derivatives but nevertheless show chromosome mobilization (Curtiss and Renshaw, 1968). The fact that plasmids are frequently transmitted in mating unaccompanied by the transfer agent suggests that this may be the major mechanism of plasmid mobilization.

Mobilization by mechanisms (1) and (2) requires genetic recombination between the transfer agent and the mobilized element. Recombination generally occurs between regions that are similar or identical in base sequence. Mechanism (3) does not require any structural similarity between the two elements. However, it may depend on the action of enzymes elaborated by the transfer agent on the mobilized element at the origin of transfer—enzymes whose degree of substrate specificity is unpredictable. Related problems will be discussed in later chapters (Chaps. 6–8, 11).

If mobilizability depends more strongly on the nature of the site of transfer origin than on the size of the element attached thereto, we would not expect all plasmids to be highly mobilizable. Whereas high mobilizability can be taken as direct evidence for extrachromosomal nature, low mobilizability does not argue strongly in favor of a chromosomal location.

VEGETATIVE SEGREGATION

If a plasmid has been added to the genome by conjugation or infection, or if it has become demonstrable through mutation of a preexisting plasmid, there may be occasional abnormal cell divisions where one daughter cell is plasmidless. There may also be occasional cells that lack the bacterial chromosome. These will be inviable. But a cell missing a nonessential plasmid will produce a colony lacking the phenotypic properties conferred by the plasmid genes.

In this case, nonchromosomal inheritance manifests itself as an apparent high mutability of some character. How can loss of a plasmid be distinguished from a high mutation rate of a chromosomal gene? Ultimately, this can be done only by some crossing procedure, which pushes the burden of proof back onto the first criterion.

This is an important point. Much confusion can result from the designation of gene blocks as plasmids solely because they can be lost from the cell. The basic criterion must always be whether they are connected to the bacterial chromosome. Operationally, this point can be established only by observation of their linkage to chromosomal markers in crosses.

In general, it is possible to distinguish loss of a block of genes from mutation of one of them if enough genes are available. However, cases are known where deletions of particular blocks occur with relatively high frequency, although the block, when present, is linked to other chromosomal genes.

Vegetative segregation in bacteria has a long history. In the 1940s, Lederberg found that some strains of *E. coli* yielded, on crossing, progeny diploid for much of the bacterial chromosome. Diploidy was manifested by rapid segregation of haploid or homozygous individuals displaying recessive characters put into the cross. Diploidy extended over most, but not all, of the known genetic map.

These diploids were the object of several years of intensive study, because it was hoped that they were intermediates in bacterial conjugation. After the analysis of Hfr $\times$ F$^-$ matings by Jacob and Wollman, it seemed more likely that diploid production reflected a structural abnormality of particular strains rather than stabilization of a normal intermediate. Interest in the diploids waned, although no real understanding had been achieved as to what they were or how they came about.

In the early 1960s, Adelberg and Jacob discovered the first F′ strains. The F agent, coming from its chromosomal location, occasionally could bring with it a section of the adjacent bacterial chromosome. The new hybrid element could spread through the population like F. Cells harboring F′, unlike the Lederberg diploids, could be "cured" of their diploidy with acridine dyes. Partial diploids could now be made, in principle, at will. A variety of strains carrying different segments of the bacterial chromosome have now been synthesized. We shall return to their significance and mode of origin in Chap. 8.

The Lederberg diploids showed none of the properties characteristic of cells carrying F′. Besides being incurable by acridine, they were not necessarily F$^+$. More recently, several other cases of partial diploidy have been reported, which resemble the Lederberg strains in both these respects. Furthermore, if these strains are made F$^+$, the "extra fragment" of bacterial genome acquires neither the high infectivity nor the acridine-curability of F. There is no sign of any abortive or incomplete F′ which could be regenerated by contact with F (Curtiss, 1964*b*; Campbell, 1965*a*).

In some instances, it has been shown that diploidy is chromosomal; in crosses, the property of being diploid for certain genes is closely linked to those genes. The clearest case of this type was one studied by Horiuchi *et al.* (1962). Diploidy of the *lac* region mapped at the *lac* locus in Hfr $\times$ F crosses. P1 transduction to *lac*$^-$ recipients caused replacement of the recipient *lac* genes with the diploid *lac* locus of the donor—showing that both copies of the region were carried by a single P1 particle.

With other diploid strains, diploidy is transferable by P1 transduction, but only one of the two allelic regions is transmitted by any one particle. Separate transfer of the two regions is also observed in

$F^+ \times F^-$ crosses. These results are compatible with either of two models: (1) Diploidy results from the presence of one set of genes as an extrachromosomal fragment, perhaps attached to some undiscovered plasmid. (2) Diploidy is chromosomal. The observed lack of linkage results from the length of the diploid region or its position on the chromosome. The pattern of spontaneous haploidization of such

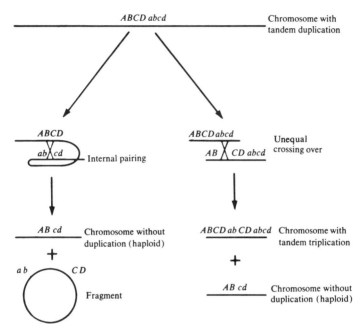

FIG. 4-1. Possible mechanisms of haploidization of a chromosome containing a tandem duplication. Only one type of recombinational event is shown in each case. The haploid segregants constitute a "polarized" set of recombinants—*ABCD, ABCd, ABcd, Abcd, abcd,* without the reciprocal classes.

Internal pairing creates a small ring of genetic material, which presumably is lost. Unequal crossing over will give one triploid individual for every haploid segregant.

strains suggests that haploidization involves recombination. The simplest picture is a tandem duplication of genetic material. As diagrammed in Fig. 4-1, haploidization could occur either by internal crossing over within one chromosome or by unequal crossing over between two chromosomes.

Until more information is available, an open mind should be kept as to which of these alternatives applies in each case.

An illuminating situation has been described by Hirota and Uchida (1964). If a cell carrying an F' agent is mated to an F⁻, no linkage is observed between the bacterial markers carried on the episome and their chromosomal homologues. However, if the F' is subjected to ^{32}P suicide, some survivors are found that are still diploid, but where the marker originally on the episome is now closely linked to its homologue on the chromosome. Apparently, once the bacterial genes have been separated from F by phosphorus decay, they can no longer replicate autonomously. The precise mechanism by which chromosomal diploids are then generated is not understood in detail.

Partial diploids, whether of the F' or the chromosomal insertion variety, raise a problem that can be phrased in two almost equivalent ways: (1) How do the diploids maintain their diploidy? (2) Why is *Escherichia coli* normally haploid? The latter question is pertinent when one considers the growth of an F⁻ cell that has just received the entire Hfr chromosome by bacterial mating. The Hfr chromosome, according to the picture we shall develop in Chap. 8, should be equivalent to a big F' which contains the whole bacterial chromosome. It should be able to reproduce either independently of, or integrated within, the chromosome of the recipient. Yet bacterial recombinants that have received the entire donor genome are typically haploid, not diploid.

Experimentally, this question has not been at all probed to its depths. The explanation could either be trivial and quantitative or interesting and qualitative. Large F's are expected to be less stable than small F's. There is more opportunity for them to recombine with the bacterial chromosome (producing either homozygosis or haploidy, depending on whether the element is free or integrated). Their infectious spread is also less efficient because of the longer transfer time required, so that cells which have lost the F are less likely to regain it by infection. It could be that a duplication the size of the whole chromosome, either in an F' or as an insertional chromosomal diploid, is so large that it cannot be maintained.

Alternatively, there could be some special points or regions of the bacterial chromosome that cannot be diploid because of their relation to cell division or to chromosomal division. Eukaryotic chromosomes, for example, may reproduce extensive duplications, but even a small duplication including the centromere creates mechanical problems that prevent its perpetuation. In the bacterial case, if two identical centromeres (or the bacterial equivalent thereof) are in the same cell, they might automatically segregate from each other at the next cell division.

VEGETATIVE SEGREGATION

If such a special region exists on the *E. coli* chromosome, it must be in some section of the map where no diploids have yet been found. Tables 4-1 and 4-2 summarize the types of partial diploids that have been well characterized in the published literature. As shown in Fig. 4-2, they cover most, but not yet all, of the bacterial chromosome.

TABLE 4-1. *Partial Diploids of* Escherichia coli *Carrying* F' *Agents*

Name of F' Agent[a]	Genetic Structure	Number Used in Fig. 4-2
F-*lac*	O-*lacI-lacZ*-F	1
F-*pro*	O-*proB*-F	2
F *lac proC*	O-*proC-lacI-lacZ*-F	3
Unnamed	O-*lacZ*-F	4
	O-*proC-lacI-lacZ*-F	5
	O-*pho-proC-lacI-lacZ*-F	6
	O-*tsx-pho-proC-lacI-lacZ*-F	7
	O-*purE-tsx-pho-proC-lacI-lacZ*-F	8
	O-*gal-purE-tsx-pho-proC-lacI-lacZ*-F	9
F *lac pro*	O-*proA-proB-lacI-lacZ*-F	10
F'-X647	O-*lacI-lacZ*-F	11
F-13	O-*phoA-phoR-tsx-purE-lacZ-lacI*-F	12
F-BB1	O-*phoA-phoR-tsx-purE-lacZ-lacI*-F	13
F'-ORF-1	O-*purE-proC-lacI-lacZ*-F	14
F'-X646	O-*proA-proB-lacI-lacZ*-F	15
F-8	O-*gal*-F	16
F-3	O-*gal*-F	17
F' *gal*	O-*supB-gal-att*λ-F	18
F-try	O-*tonB-trp-cysB*-(*colV*)-(*colB*)-F	19
F-14	O-*arg-metB-met E-ilv*-F	20
F-15	O-*thyA-argA*-(*galR*)-F	21
F-16	O-*argR-argG-serA-lysA-thyA-galR*-F	22
F-32	O-*dsdA-aroC-purC*-(*gua*)-F	23
Unnamed[a]	O-*argG-argR-malA-strA-xyl-argE*-F	24

[a] Taken from Scaife (1967) with nomenclature modified to be in accord with Taylor and Trotter (1967). The last strain has been added to Scaife's list. Though its history and origin are a little different from those of the other strains in the table, this element seems to have all the properties of an F' (Maas and Clark, 1964).

Two other methods of determining the extrachromosomal nature of an element are listed below. Both are formally "vegetative segregation" tests, but each has special attributes that warrant separate consideration.

TABLE 4-2. *Other Partial Diploid Strains*[a]

Strain	Markers Present in Diploid Condition	Origin	Properties	Reference
E104[b]	*lacI lacZ lacY*	Selection for hyperproduction of β-galactosidase	Chromosomal, linked to *lac* region	Novick and Horiuchi (1961)
R688[b]	*gal att*λ *bio uvrB chlA*	Cross Hfr *gal*$^+$ × F$^-$ *gal*$^-$		Campbell (1965a)
X98[b]	*ara valS leu azi proA proB*	Cross Hfr *Cpro*$^+$ × F$^-$ *pro*$^-$		Curtiss (1964b)
Het[b]	*xyl met leu tonA lac tsx trp*	Spontaneous, detected by mutation to T1 resistance		Lederberg *et al.* (1951)

[a] In addition to the strains listed, partial diploidy for limited regions can result from incorporation of these regions into specialized transducing phages (see Chap. 8).
[b] In Fig. 4-2, these strains are coded as follows: 25 = E104; 26 = R688; 27 = X98; 28 = Het.

Curing. The rate of vegetative segregation of many plasmids, including the F agent, is greatly enhanced by exposure to acridine or certain metal ions, or by thymine starvation. The mechanism of these "curing" effects is unknown. Not all plasmids are cured with equal ease. Removal of chromosomal elements by these agents has not been reported. Some of them (the acridines) induce in bacteriophage small

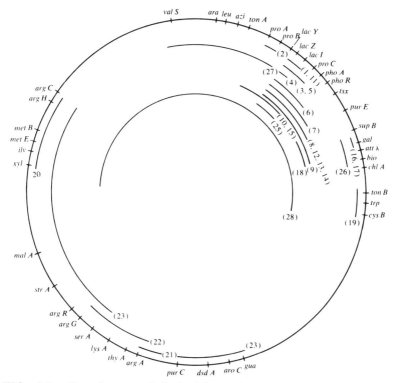

FIG. 4-2. Genetic map of *Escherichia coli* showing the extent of the partial diploids listed in Table 4-1. For each strain are shown the genes known to be diploid. The exact limits of diploidy are unknown in many cases. [After Taylor and Trotter (1967). Used by permission of the authors.]

"deletion" mutations of single nucleotides. Literally, this is removal of a chromosomal element, but at a rather different level than the loss of a large complex genetic structure. Demonstrated curing is therefore good presumptive evidence for a nonchromosomal element. Failure to demonstrate such an effect leaves the question of location entirely open. Chromosomal elements such as prophage *can* be cured by irradiation, and by superinfection with related phages (see Chap. 7).

REPLICATION-DEFICIENT MUTANTS

According to present ideas on the mechanism and control of DNA replication (discussed in Chap. 10), the ability of an extrachromosomal element to replicate should depend on its ability to synthesize some specific initiator substance(s). A chromosomal element should not show this dependency. However, a breakdown in the regulatory circuitry of an integrated element (caused, for example, by heating a mutant with a thermolabile repressor) might cause excision of the element from the chromosome and subsequent loss from the cell. Heat-sensitive mutants, in which plasmid replication fails at high temperature, have been isolated for known extrachromosomal elements such as the F factor (Jacob *et al.,* 1963).

STAPHYLOCOCCAL PLASMIDS

The best-studied plasmids of the Enterobacteriaceae—the colicinogeny and the R agents—have been considered in this and the previous chapter. For comparison, we may consider an interesting case of nonchromosomal inheritance in staphylococcus (Novick, 1967). Penicillinase production is determined by an element that can be either present or absent but which, when present, segregates rapidly. The element contains, in addition to the structural gene for penicillinase and a regulator gene for the same enzyme, independent genetic determinants for resistance to several different inorganic ions and to the antibiotic erythromycin. These determinants can be placed on a linear genetic map by three independent criteria: (1) Transduction between strains carrying genetically marked elements. (2) Segregation of haploid types from strains with two elements. Such "diploid" strains occur only if the two elements in question differ in "compatibility." Two compatibility types have ben found in different isolates from nature. Compatibility is ascribed to a genetic locus *mc* that can be mapped on the element. (3) Deletion mapping. All deletions thus far encountered have been terminal (starting from either end of the map). The *mc* locus is never deleted.

That segregation of the element reflects a physical loss of genetic material is suggested by studies on ultraviolet inactivation of the transducing ability of a lysate made on bacteria harboring the element and assayed on segregants that have lost it. The UV sensitivity of the element is large and equal to that of the individual markers within it—

as though the whole element must be introduced for transduction of any part to occur. Replication-deficient mutants of the plasmid have been found. The extrachromosomal nature of this element seems as well established as is possible in an organism where nothing much is known of the chromosome itself.

The possibility is open that the bacterial genome normally comprises many small linkage structures in addition to the single major one usually studied, and that mutations in the others are frequently either unobserved or misinterpreted. The possible mechanism by which orderly segregation of such units might be achieved during cell division will be treated in Chap. 10.

PLAN FOR REMAINDER OF BOOK

This concludes our taxonomic survey of episomes and related elements in bacteria. The following chapters will describe experimental studies of various problems raised by episomes, and the integration of these findings into the conceptual framework of modern biology. In choosing material to illustrate problems and principles, I have frequently selected the temperate bacteriophages, especially lambda. This is partly due to my own experience with this phage, but also to the amount of information available on the internal genetic and physical structure of the element itself.

A brief description of the application of these findings to eukaryotes, as well as of elements bearing some similarity to bacterial episomes in higher forms, will be given in the final chapter.

5

MODE OF CHROMOSOMAL ATTACHMENT

Several years ago, during the course of writing a review on episomes, I proposed a model for the attachment of prophages to bacterial chromosomes; or, more generally, of episomes to chromosomes. According to this model, the episome must be or become circular. It then behaves like a "ring chromosome" in classical genetics. If it crosses over with another chromosome, the result is that the genes of the ring become inserted linearly into the continuity of the other chromosome (see Fig. 5-1). At the time I proposed the model, I did not believe it was correct. It was simply the result of my best efforts to understand the genetic data then available while avoiding the distasteful notion that the episome was stuck onto the outside of the chromosome. The facts as they stood then seemed to speak as much against the model as for it.

I began to take the picture seriously only after it was found that the genetic constitution of segregants from bacteria doubly lysogenic for phage lambda was precisely that predicted (Campbell, 1963a). This was by no means the most critical test of the model possible. We just happened to accumulate the data during work with these strains, and it impressed me that the results should fit so well the attractive simple picture I had recently drawn.

Since 1963, various laboratories have supplied convincing evidence that the mode of attachment shown in Fig. 5-1 is correct. The prophage is inserted into the continuity of the bacterial chromosome, and the prophage gene order is a cyclic permutation of the order in the free phage, as in the last line of Fig. 5-1. Almost all the experimental evidence comes from phage lambda and its relatives.

The results leading to this conclusion have all consisted, in one manner or another, of constructing a map of the lysogenic chromosome. The first work in this direction was done by Jacob and Wollman in 1957. They made bacterial crosses between lysogenic parents and looked at recombinants between the bacterial markers *gal* and *21*, which bracket the prophage. The prophages of the two parents were marked by the three mutations *m5, co,* and *mi.* Their results are shown in Table 5-1.

It is easier to interpret their data in retrospect than it was at the time. If one assumes that the gene order of the prophage is the same as that of the free phage,[1] it appears that prophage recombination follows its own rules. The pattern of recombination seems to be in-

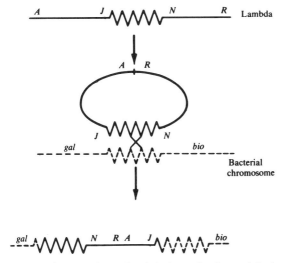

FIG. 5-1. Integration of phage lambda into the bacterial chromosome.

fluenced by association with the bacterial chromosome, but too independent of it to be compatible with insertion. Jacob and Wollman made this assumption, which is certainly a reasonable one. In an entirely new kind of experiment, some internal control is desirable. The most elementary control is that the three prophage markers should remain in their proper order.

The only person I know of who interpreted their result in a different light was Enrico Calef. Calef noticed that Jacob and Wollman's result was compatible with linear insertion provided it was assumed

[1] By "gene order of the free phage" we mean that indicated in Fig. 2-1. This map is based on frequencies of recombinant types liberated by cells after mixed infection. The correspondence between this map and the DNA molecule of the free phage particle is well established (see Chap. 2).

that the gene order was different in the prophage than it was in the free phage. His approach clearly simplified the problem because it replaced two questions with one. Instead of postulating (1) an unknown mode of prophage attachment and (2) an unknown effect of attachment on prophage recombination, Calef proposed that the mode of attachment is simple insertion and that the recombination pattern is that characteristic of ordinary recombination between linear structures.

TABLE 5-1. *Recombination of Lambda Prophages in Crosses Between Two Lysogenic Parents*

Cross: Hfr $gal^+(\lambda m5\ c^+\ mi^+)(21)^-str\text{-}s \times F^-gal^-(\lambda m5^+\ c\ mi)(21)^+str\text{-}r$
Selection: $gal^+\ str\text{-}r$

Types of prophages found in recombinants		Number found	
		In the class $Gal^+(21)^+$	In the class $Gal^+(21)^-$
Parental types	$m5c^+mi^+$ (Hfr type)	276	1
	$m5^+c\ mi$ (F⁻ type)	57	127
Recombinants	$m5^+\ c^+\ mi^+$	12	4
	$m5\ c\ mi$	1	0
	$m5\ c^+\ mi$	3	2
	$m5^+\ c\ mi^+$	1	0
	$m5\ c\ mi^+$	2	0
	$m5^+\ c^+mi$	6	1
Double lysogens	$m5\ c^+\ mi^+$ and $m5^+\ c\ mi$	3	0
	$m5^+\ c^+\ mi$ and $m5^+\ c^+\ mi^+$	1	0
	$m5\ c^+\ mi$ and $m5\ c^+\ mi^+$	1	1

Location of markers on the lambda map

```
  A         J             N         R
  ──────────────WWW─────────────────
              m5          c        mi
```

SOURCE: Jacob and Wollman (1957.)

He left the one question of how a change in gene order might come about during prophage attachment.

If Jacob and Wollman's data are interpreted as they were by Calef, they fit the insertion hypothesis reasonably well. The analysis is shown in Table 5-2. In the first cross, where an odd number of crossovers was selected, the most common classes are the four single crossover types (of which only the two representing recombination within the prophage are really critical). In the second cross, an even number of crossovers was selected. The four recombinant types found are all doubles; the one possible quadruple type does not appear.

The numbers were small, but they were suggestive enough to encourage Calef to extend Jacob and Wollman's observations further —with the introduction of a few technical improvements that facilitated examination of a larger number of recombinants. The first crosses employed the three phage markers h, $cIII$, and mi. Selection

TABLE 5-2. *Analysis of Data in Table 5-1*

Crossovers according to prophage map: *gal I c II mi III m5 IV (21)*

	Number of occurrences
Class (gal^+) $(21)^+$ of Table 5-1	
Single I	57
Single II	6
Single III	12
Single IV	276
Triple I II III	1
Triple I II IV	2
Triple I III IV	1
Triple II III IV	3
Class $(gal^+)(21)^-$ of Table 5-1	
Noncrossover	127
Double I II	0
Double I III	0
Double I IV	1
Double II III	2
Double II IV	1
Double III IV	4
Quadruple I II III IV	0

was made for an odd number of crossovers between *gal* and *trp*. The data fit the order *gal cIII mi h trp*. As *m5* is close to *h*, this result was essentially the same as Jacob and Wollman's. Calef later extended these observations to experiments where an even number of crossovers were selected, and also to crosses involving other markers. All his results have corroborated the prophage map shown in Fig. 5-1.

In 1963, I reported on the segregation patterns from double lysogens carrying one lambda prophage and one lambda *dg* prophage. These results will be discussed in Chap. 12. Later, crosses between lambda prophage and irradiated lambda *dg* phage were performed (Campbell, 1964). These experiments did not provide direct evidence on the relationship between prophage and bacterial chromosome, but they were highly indicative that the gene order within the prophage was that shown in Fig. 5-1.

In 1965, Rothman mapped the lambda prophage and bracketing bacterial markers (in this case *gal* and *bio*) by P1 transduction. She found that prophage attachment "stretches" the bacterial genetic map in the *gal bio* region. The cotransduction frequency of *bio* and *gal* dropped from 47 percent to 7 percent when both donor and recipient were made lysogenic for lambda.

The significance of this sort of stretching experiment has frequently been exaggerated. The classical genetic literature is replete with examples where the recombination frequency within an interval is strongly influenced by the nature of adjacent regions—proximity to the centromere, for example. It is far more convincing to look at the relative proportions of the different recombinants that occur in crosses between lysogens than it is to compare the lysogenized chromosome with its nonlysogenized counterpart. Indeed, I would be surprised if addition of a lambda-sized piece of DNA between *gal* and *bio* failed to affect recombination in this region, whether or not the addition comprised a physical insertion into the chromosome. There is also an anomaly in Rothman's map-stretching result, in that much of the additional recombination ascribable to the presence of prophage seems to take place between *gal* and lambda rather than within the prophage.

An additional complication appeared in Rothman's data. Crossing a gal^+ bio^+ lysogenic donor by a gal^- bio^- lysogenic recipient, one can select either for gal^+ transductants or for bio^+ transductants. The frequency of joint transduction of *gal* and lambda is low, but the major classes of recombinant prophages found are those that would be single crossovers according to Fig. 5-1. When bio^+ transductants are selected, the joint transduction frequency of lambda is high, but very few recombinant prophages are found. The small number of observed recombinants do not fit any simple pattern, although the phage marker most closely linked to *bio* is that expected from Fig. 5-1.

All in all, Rothman's results supported the insertion model, but they uncovered some complications which are still unresolved. The asymmetry of the result depending on the selective marker employed is still unexplained. Maybe this asymmetry reflects some peculiarity of P1 transduction with respect either to the *gal bio* region itself or to lambda prophage. Or perhaps the *gal bio* recipients used differ from the gal^+ bio^+ donor stock by structural rearrangements in the *gal-lambda-bio* region. At any rate, the asymmetry observed by Rothman was not found in later studies by Zavada and Calef (1968) on lysogenization of *E. coli* B by lambda and lambda *b2*.

The most definitive argument for prophage insertion comes from the work of Franklin *et al.* (1965) and Gratia (1966). Their experiments employed a relative of lambda, phage 80. Prophage 80 is located on the bacterial chromosome close to the *tonB* gene that causes

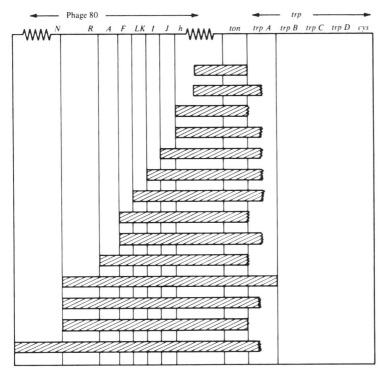

FIG. 5-2. Some of the deletion mutants isolated in the *trp-ton*-phage 80 region. Taken from data of Franklin *et al.* (1965). Cross-hatched area represents deleted region. Where the right end of the area is drawn with a sawtooth line, the exact extent to which the deletion penetrates the *trp* region was not measured.

sensitivity to phage T1. The rare T1-resistant mutants arising in the bacterial population frequently prove to be deletion mutants lacking the surface receptor for phage T1. The deletions sometimes penetrate the *trp* operon, which is on the other side of *tonB* from prophage 80. Franklin *et al.* showed that such deletions can also penetrate the prophage (Fig. 5-2). The use of deletions for mapping is free from many quantitative problems that complicate recombinational analysis. As emphasized by Benzer (1959), the compatibility of a collection of deletion mutants with a unique linear arrangement of genes is a strong argument that such a linear arrangement really exists. In the case at hand, the argument implies that the *trp* operon, the *tonB* locus, and prophage 80 all constitute one linear system.

In these experiments, Franklin *et al.* did not use wild-type phage 80 but rather a hybrid between lambda and phage 80. The mutants used in the mapping were all derived from the "lambda" part of the hybrid phage. The fact that the data show not only insertion but also the same cyclic permutation found for lambda thus completely accords with other work. Deletion mapping of the lambda prophage itself has more recently been accomplished (Shapiro, 1967; Adhya *et al.*, 1968), with the expected results. Phage 80, like lambda, increases the genetic map distance between bracketing bacterial markers (Signer, 1966). The P22 phage of *Salmonella typhimurium* shows a circular map in ordinary crosses (Gough and Levine, 1968), but the prophage map is a unique linear derivative of that circular map (Smith and Levine, 1965).

To this may be added a large body of information on the gene content of transducing phages. This is the inverse of the "deletion mapping" approach described above. Instead of looking at what remains in the chromosome, we look at the region that has fallen out. Ordinarily, such chromosomal fragments are unrecoverable. However, a fragment that contains part of the prophage as well as some bacterial genes may survive, thanks to its ability to multiply as a phage. A detailed description of transducing phages will be given in Chap. 8. From the gene content of different transducing phages, the map order of both prophage and bacterial markers can be determined. It was because this method indicated the same permuted order for the prophage as that deduced by Calef from bacterial crosses that I was attracted to the model of Fig. 5-1 in the first place.

A scientific hypothesis is usually either accepted or rejected. Today's hypothesis, if it does not prove obviously incorrect, becomes tomorrow's dogma. The hypothesis schematized in Fig. 5-1 was proposed in a search for a simple connection between certain observations on a complex system. For several years it was necessary to gaze very optimistically at those selected facts that seemed to fit our picture, and to hope that other less agreeable findings would somehow disappear from view.

There is frequently a phase in the development of an idea when this optimistic approach is desirable. Textbooks on the scientific method generally instruct us to take the facts as "given" and to restrict our hypotheses to those that fit the facts. The textbook approach is logical in principle. It fails in practice (except perhaps in some areas of physical science where the operational bases for collecting "facts" are exceedingly well defined) for a reason that concerns scientists rather than science. In the absence of an explicit model, it is hard to collect the right facts, and to be correct about all the minor details which may turn out to be critical to a model constructed later on. The approach that has proven fruitful in the present case has not been to ask "What model can accommodate all the facts?"—but rather, first to consider the major aspects of the situation, to ask whether these might fit into some simple picture, and then to examine very critically any "fact" that would require the adoption of a complex model in place of a simpler one.

During the last few years, some of the facts that seemed to contradict our simple picture have been disposed of. For example, before the work of Rothman, it was thought that the biotin gene was between *gal* and lambda rather than beyond lambda on the bacterial chromosome. This underscores the virtual impossibility that any data will be correct in all details—especially on details that seemed of no special importance at the time the work was done.

Other facts that seemed to speak against this model came from quantitative studies of lysogenization frequencies. These have turned out not to refute the model but to complicate it. The complications are not at the genetic but rather at the physiological level. The direction in which these facts are leading us will be described in Chap. 6.

How should we consider our model in light of present information? In the period after the model was proposed, the need was for optimism. The model has now achieved fairly general acceptance. The need now is for criticism. Do the facts recounted in this chapter provide convincing evidence that the model is correct? What remains to be done to make the argument even tighter?

What is under discussion is not the entire model of Fig. 5-1, but only one of its consequences—the mode of prophage attachment. This mode could arise equally well if, for example, the bacterial chromosome were first to recombine with the linear phage chromosome, and the free ends of the resulting linear structure were then to join together. The question of whether end joining precedes or follows integration must be argued independently on the basis of different evidence.

The asymmetry in Rothman's mapping data is a case in point. The accessory assumptions required here concern the process of genetic assortment itself and cannot be explained by reference to physiological complications. Rothman's own evaluation of the results was very temperate. She concluded that her data could fit an insertion model as well as any other model. As an optimist, I was pleased that her results with *gal* transduction gave the predicted order, and that the biotin gene was where I wanted it to be. As a critic, I must admit that any argument to give the *gal* transduction data preference over the *bio* transduction data only indicates a prejudice in favor of the model. It is clear that there are some unresolved complications, which may lead to the discovery of some interesting and relevant phenomena, but which preclude using her results as the basis of a firm argument in favor of insertion as depicted in Fig. 5-1.

Franklin's experiment impressed me as quite convincing. Topological mapping provides a strong case both for the linearity of the structure mapped and for the order of genes along that linear structure. Especially taken together with the data on transducing phages (Chap. 8), the results allow no simple alternative to the insertion model.

If we compare the evidence that the prophage is inserted into the bacterial chromosome with that showing that any authentic bacterial region (say, the *lac* operon or the *trp* region) is itself colinear with the rest of the genome, I think we must accept that the former is about as good as the latter. There are some complications and anomalies, but no more than are found in bacterial genetics in general.

Therefore, while I would urge a continued critical approach to the basic hypothesis, I think that the model of Fig. 5-1 has passed enough tests so that it should tentatively be adopted. By this I mean explicitly that wherever an experimental result can be accommodated to the model only by the introduction of additional accessory hypotheses or unknown elements, we should make what hypotheses and postulates are necessary to fit the results to the model and hope that these may lead to the discovery of something new or interesting. This will be our approach throughout the remainder of the book.

Of all the evidence that prevented immediate adoption of this model in 1962, most of it either has proven erroneous or else can be explained by introducing new elements into the model in ways that do not negate its essential simplicity. One case that has instead just faded away without reinvestigation is phage 18. Prophage 18 maps between two methionine markers, but does not increase the map distance between them. And it has the curious property that, in time of entry experiments, the prophage enters after a given methionine marker no matter what the polarity of the Hfr donor strain, as though it were synapsed parallel to a region of the bacterial chromosome including this marker.

As indicated above, I do not regard the map-stretching experiments as critical. Moreover, as the size of the phage is unknown, there is no way to predict the expected stretching. The time of entry experiments must require some special explanation, but it is difficult to guess at the right one. This phage does not seem to have been investigated further, and the original observations were never published in full detail. The available facts could be explained if the prophage had more than one possible site of attachment and happened to have added at different sites in different strains. In my opinion, information presently available does not justify any definite conclusion regarding the mode of attachment used by phage 18. And, whereas much of the interest in our model has resulted from its applicability to entities as apparently diverse as phage lambda and the fertility agent, we should retain an open mind as to possible modes of attachment in other episomes.

MULTIPLE ATTACHMENT SITES AND ALTERNATIVE MODES OF ATTACHMENT

As regards specificity of attachment, known episomes range from phages lambda and P22, with highly efficient integration at unique chromosomal sites, through the F factor and phage mu (Taylor, 1963), which add at low frequency at a variety of sites. Lambda addition is easier to study because the site is unique, and because the genetics of the free episome are better known. The comparative biology of episomal attachment includes several intermediates between lambda and F.

Until recently, lambda attachment seemed completely site specific. In all lysogens examined, lambda mapped between *gal* and *bio*. With the availability of mutants defective in normal prophage integration (see Chap. 6), it could be shown that, very rarely, addition at other sites can occur (Gottesman and Yarmolinsky, 1968). The same appears to be true for phage P22 (Smith, 1968).

Infrequent attachment at secondary sites has been known for some years in phage P2. Lysogenization at secondary sites (of which at least nine have been distinguished) happens with low frequency when a P2 lysogen is superinfected with a mutant of P2. Phage coming from secondary-site lysogens differs from wild-type P2 in that it has lost the strong "site preference" that characterized the original strain. Six (1966) suggested that the two sites differ from each other in the manner indicated in Fig. 5-3. The essential feature is that integration and excision can occur at different points within the attachment region. If host and phage chromosome are nonidentical for some part(s) of this region, the phage that comes out can be different from the one that went in. Whether some single large stretch of the region is completely dissimilar between the two (as an extreme interpretation of Fig. 5-3 might suggest) cannot be decided. A corollary of Six's idea is that phage detachment from site II lysogens should some-

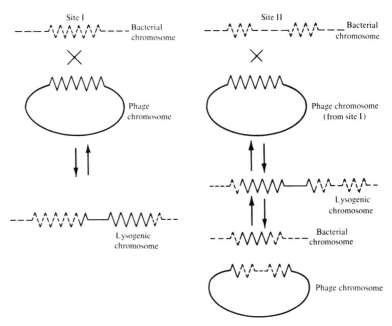

FIG. 5-3. Attachment of phage P2 at two different chromosomal sites. In attachment of standard P2 at site I, phage and bacterial attachment regions are sufficiently similar to be treated as identical. This is exactly equivalent to lambda attachment as shown in Fig. 5-1. At site II, the bacterial attachment region is interrupted by a section that is dissimilar to the attachment region of phage coming from site I. Depending on where the crossover occurs in detachment, site II lysogens can give either standard P2 or a new P2 whose attachment region includes that section of bacterial genes which is intercalated into site II. This new P2 shows a decreased preference for site I as compared with standard P2. [Modified from Six (1966). Used by permission of the author.]

times produce bacteria in which site II has become structurally identical to site I.

The F agent can attach to the chromosome at many different sites (Fig. 1-3). That the number of potential sites is probably much larger than indicated in Fig. 1-3 has been shown by some very ingenious experiments with a temperature-sensitive F *lac* episome (Beckwith *et al.*,

1966). The origin and properties of this episome will be discussed in more detail in Chaps. 8 and 10. This is an F particle that has incorporated the *lac* gene of the host and that is unable to replicate autonomously at high temperature. Those cells capable of producing lac^+ colonies at high temperature are those in which F has become incorporated into the chromosome.

F lac is usually inserted in the *lac* region, which offers the greatest homology to the episome. If the chromosome carries a deletion of the *lac* region, insertion must occur elsewhere although it might still occur through the *lac* moiety of the episome rather than the F moiety. A variety of different Hfr strains were obtained by this method, some of which represent insertion at sites already observed with F itself. In some such Hfr strains, *lac* behaves as the most proximal marker in mating experiments; and in others, as the most distal. If insertion follows the model of Fig. 5-1, this result implies that the integration crossover can involve different sites on the episome as well as on the chromosome.

If insertion happens within or near a bacterial gene, the function of that gene can be inactivated (Curtiss, 1964a). If insertion can happen within any gene, then it should be possible to select for specific insertion at those genes whose inactive state confers a selective advantage. This is the case for phage-resistance loci. The normal gene causes synthesis of a receptor on the cell surface to which, say, T6 phage can attach. It thus renders the cell phage-sensitive. Inactivation or deletion of this gene causes phage resistance.

Insertions of F can be selected at the genes governing sensitivity to various phages. This indicates that the number of possible insertion sites for F (or at least for this *F lac*) must be very large. While some sites are preferred, insertion can happen almost anywhere, in or near almost any gene. An episome that resembles F in this respect is the "mutator phage" mu (Taylor, 1963). Here also a change in gene function at the site of prophage integration is frequent.

Little is known yet as to the detailed mechanism of these addition events. Were addition to proceed as in Fig. 5-1 or Fig. 5-3, the episome might inactivate a gene either by interrupting its base sequence or by deranging its control mechanisms. Such inactivation should be reversible. Detachment of the episome should sometimes proceed by exact reversal of incorporation, leaving the original chromosome intact. In fact, however, inactivation is commonly irreversible. Even though the episome is able to detach, the host chromosome is permanently altered somewhere in the vicinity of the attachment site.

One of the basic ideas of the episome concept has been that an episome adds to the chromosome rather than replacing a portion of it. Our model is designed to embody this idea. Two circular structures cross over and become one, but nothing is lost in the process. This should be so no matter how limited the similarity of base sequence in the crossover region, or how many further complications we introduce into Fig. 5-3 as to the disposition of these regions.

If the observed "mutations" accompanying episome attachment should turn out to be deletions, genuinely alternative models for attachment will need consideration. On the other hand, some highly mutagenic feature of the recombination process itself in the (chemically) immediate vicinity of the recombinational event would likewise fit the observations. The cases studied have been selected in such a manner that nonmutagenic incorporations would not be found.

The idea that F, like lambda, really is inserted into the chromosome derives primarily from studies on F′ strains, to be detailed in Chap. 8. Even there, the evidence is by no means compelling. Attachment of F does increase the recombination frequency between bacterial markers (Pittard, 1965).

It seems reasonable to suppose that F sometimes is inserted by a single crossover between episome and chromosome, and that this commonly involves similar base sequences of the two, where they are available. There may also be other possible mechanisms that are sometimes used. Some of these may have more in common with the "abnormal detachment" reactions described in Chap. 8 than with the normal detachment mechanism covered in Chaps. 6 and 7.

It would be a mistake to force all of these observations into a common interpretation based on the lambda work. We can be confident that the individual mechanistic steps of prophage integration will find counterparts in other systems; but this does not mean that all episomes need employ the same tricks. The whole comparative picture, when available, should make a fascinating story. The most immediate need seems to be for additional studies on the fine structure genetics of mutations associated with episome attachment. Other studies on the mode of attachment must await advances in the genetics of F and mu themselves, and hopefully, of physical techniques sufficiently refined to render genetics unnecessary.

6

MECHANISM OF CHROMOSOMAL ATTACHMENT

Assuming the mode of attachment depicted in Fig. 5-1, we can examine the mechanism of attachment. Is the order of events really that shown in Fig. 5-1, and what can be said in detail about the nature of the individual steps?

Figure 5-1 indicates two basic events in prophage attachment: (1) The ends of the phage chromosome become joined. If this happens first, the result is a circular phage chromosome. (2) Reciprocal recombination between host and phage chromosome occurs in the *att* region, thus attaching the phage chromosome onto the bacterial chromosome. If this happens first, the result is a linear bacterial chromosome with phage genes at both ends.

The biological and physical evidence concerning circularization and the mechanism of end joining will be covered in Chap. 11. We shall concern ourselves here with the recombinational event. There are still many gaps in our information, but some aspects are becoming clearer.

PHYSICAL NATURE OF THE RECOMBINATIONAL EVENT

The diagram of Fig. 5-1 is based on genetic rather than physical information. It does not automatically imply that the DNA molecule of an infecting particle (rather than a copy thereof) can become physically incorporated into the bacterial chromosome. That genetic recombinants can arise by breaking and joining of parental molecules is well established. Prophage insertion is a special kind of recombination process, for which the point had to be independently established. The best present evidence is that the DNA molecule of the infecting particle can be incorporated bodily into the bacterial chromosomes, as expected if insertion occurs by breaking and joining (Hoffman and Rubenstein, 1968).

As in the case of recombination in general, this shows that some lysogenizations occur by breaking and joining, but does not preclude that others (perhaps most) come about during DNA synthesis. We shall assume for the purposes of discussion that reciprocal breakage and joining is the only mechanism of lysogenization.

Earlier evidence indicated that the DNA molecule of the infecting particle usually is not the molecule that is inserted into the chromosome to become prophage. The simplest explanation is that autonomous phage replication frequently precedes chromosomal attachment. The DNA molecule inserted into the chromosome is then usually a copy of the molecule that infected the cell, even if the act of insertion occurs by breaking and joining.

The best indication that autonomous replication can precede insertion comes from Bertani's (1962) work with phage P2. He showed that the ratio of double to single lysogeny was independent of the multiplicity of infection—which is easily understood only if a single infecting particle can multiply before lysogenizing. Other arguments have been based on the multiplicity dependence of lysogenization under conditions where phage multiplication is inhibited (Brooks, 1965). Under conditions where multiplication is allowed, the frequency of lysogenization by phage lambda is not strongly multiplicity dependent. This is readily understood if, under the latter conditions, a large pool of autonomous phage accumulates, effectively creating a high multiplicity. Unfortunately, the agents that block multiplication (phage-specific immunity or conditional lethal mutations) may also interfere with the attachment mechanism itself, so that no rigorous argument can be made from these observations.

NATURE AND LOCATION OF THE CROSSOVER REGION

If there is a region in which crossover between phage and bacterium can occur, it is clear that the bacterial region must be located at the prophage attachment site. The phage region must lie at that point on the phage chromosome where a crossover would produce the observed permutation of markers—i.e., at the locus designated *att* in Fig. 2-1.

Different phages attach at different sites. This implies a specific recognition of the phage and bacterial regions concerned. Either the two regions recognize each other (as in homologous pairing), or a third system recognizes both. Genetic variation within the phage recognition region might then have two possible consequences: (1) a change of specificity of prophage location, so that the phage now attaches to a different bacterial site and (2) loss of ability by the phage to attach to the bacterial chromosome at all. The second type of change has been observed. The first has not (which is hardly surprising, in that it involves creation of a new specificity as well as destruction of an old one). But it can still be studied genetically because some naturally occurring phages with different attachment sites can recombine with each other. If the specificity of prophage attachment lies in the region designated as *att,* then this property should assort in such crosses in a manner consistent with this location.

The data on this point are not very strong. The main difficulty is that the critical crosses are between essentially different species whose chromosomes present only occasional isolated stretches of similar base sequence. It is hard to know how many of the potential recombinants may be lost due to inviability. Some facts are worth recounting, however.

The first problem explored was whether the determinants of prophage localization were separable by recombination from those controlling specificity of superinfection immunity. Crosses between phages lambda and 434 at one time indicated that these two properties had a common genetic determinant. However, subsequent work has not substantiated this interpretation. In fact, it is quite possible that these two phages really attach at the same chromosomal site. The idea of two distinct sites comes mainly from differences in frequency of zygotic induction and in degree of linkage with the other markers in bacterial crosses—differences that might depend on the length and composition of the prophage as well as its location. No bacterial genes are known to lie between these two sites. The most exacting tests should be three-factor crosses to observe directly the allelism or non-allelism of the two prophages; however, these present technical difficulties, such as transfer induction, which render their interpretation quite difficult.

For phage 21 and 80, whose attachment sites are far from that of lambda, attachment site specificity is clearly determined by a locus different from the immunity region. Lambda-80 crosses give some hope that this locus is close to h, as the *att* region seems to be, but a precise mapping of attachment specificity is not yet available.

A variant of phage P2 is known that can occupy the same chromosomal sites as P2, but which has a different immunity specificity. This indicates again that the two characteristics may be due to different determinants.

Loss of the *att* region through a deletion mutation should produce phage particles unable to attach. Whether or not they would be viable depends on whether the deletion includes, either within or outside of the actual recognition region, any genes necessary for phage growth.

In 1961, Grete Kellenberger discovered a lambda mutant with the expected properties. It maps between *h* and *cIII*, and is unable to attach to the bacterial chromosome in single infection. Its buoyant density in a CsCl gradient is less than that of lambda, indicating a deletion of roughly 17 percent of the lambda DNA. It makes normal plaques and therefore is missing no essential genes. It was called *b2* (*b* for buoyant).[1]

Sedimentation studies and contour length measurements have confirmed that the density difference does indeed measure DNA content. In crosses between genetically marked *b2* stocks, the linkage between *c* and *h* is tighter than in wild-type lambda, suggesting that the missing DNA has been taken from the region in which the mutation maps (Jordan, 1965). I really should not use this "shrinking" argument after my previous remarks about "stretching" effects in lysogenization; but anyway, shrinking is observed. Electron microscopy of hybrid DNA molecules has since confirmed that *b2* indeed has a deletion, located in the expected place (see Fig. 2-5).

The role of the $b2^+$ allele in prophage attachment might be a structural function, requiring direct contact between this region and the bacterial chromosome. This is what we expect if the mutation is really a deletion of the *att* site. Alternatively, $b2^+$ might instead have a physiological function—say, the elaboration into the cytoplasm of a diffusible "attachment principle." Zichichi and Kellenberger showed that $b2^+$ phage can effect attachment of lambda *b2* in a cell simultaneously infected with both phages. However, *b2* phage was never found on the chromosome alone. Wherever it was attached, $b2^+$ was attached also. When a double lysogen loses one of its two prophages, it is always *b2* that is lost and $b2^+$ that is retained.

None of these findings discriminates between a structural vs. a physiological function for $b2^+$. If the function is structural, the formation of double lysogens with *b2* and $b2^+$ is explained by the fact that, as diagrammed in Fig. 6-1, the integrated prophage provides a region of chromosomal homology that substitutes for the normal one lacking in the *b2* mutant. If it is physiological, then $b2^+$ would act by produc-

[1] The *b2* mutation studied by Kellenberger was apparently present in the "reference type" lambda of Kaiser (see Chap. 2).

ing an attachment principle that acts equally on *b2* and *b2*⁺ genomes in the common cytoplasm. Depending on whether the function is required for the act of attachment or for the stability of the attached state, a single lysogen for *b2* might or might not be stable once formed.

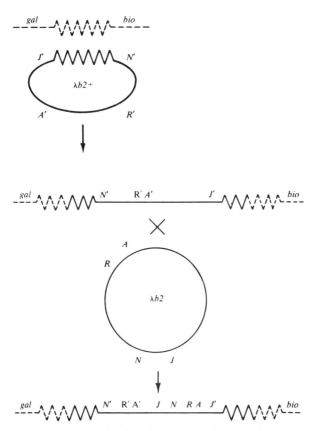

FIG. 6-1. Integration of lambda *b2* (depicted as missing the structural region involved in prophage attachment) within a *b2*⁺ prophage in a mixedly infected cell. This illustrates the way that lambda *b2*⁺ might help lambda *b2* to lysogenize through a purely structural interaction. Note that prophage loss by internal pairing can result only in a lysogen for lambda *b2*⁺, never one for lambda *b2* (cf. Fig. 4-1).

To discriminate between these two possibilities, use was made of a derivative of *Escherichia coli* K-12 that is diploid for the prophage attachment site and adjacent genes. If the way that *b2*⁺ helps *b2* to be a prophage is through some physiological function, then the two pro-

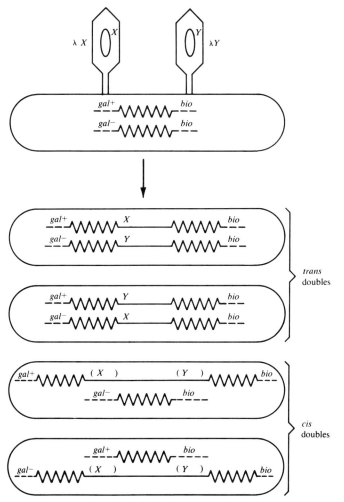

FIG. 6-2. Illustration of the two types of double lysogen obtainable from bacteria diploid for the lambda attachment site. Permutations of gene order within *cis* doubles are not illustrated.

phages should be able to form a *trans* double lysogen, with the two prophages at different sites (see Fig. 6-2). On the other hand, a structural function of the type shown in Fig. 6-1 would inevitably form only *cis* doubles. In fact, *trans* doubles are never found in $b2^+ \times b2$ infections, although they are more common than cis doubles in $b2^+ \times b2^+$ controls (Campbell, 1965*b*).

This indicates that the *b2* region plays a structural role in prophage attachment. Whether it might play a physiological role as well cannot be ascertained from the experiment.

The *b2* mutant thus has the properties expected if the region designated *att* in Fig. 2-1 is deleted. More recent evidence suggests that the crossover actually does not take place within the *b2* region but very close to it. In all our diagrams, *att* is placed at the actual point of integration, not within neighboring regions that play a crucial role in the process.

PHYSICAL STUDIES OF BASE SEQUENCE SIMILARITIES

The preceding evidence indicates that a specific region of the lambda chromosome located between *c* and *h* plays a structural role in recognition of the proper chromosomal site. Integration takes place by a crossover in the same general part of the phage chromosome. This region, in which the integration crossover occurs, we designate as *att*. The fact that we have used the same symbol for the integration sites on the phage chromosome and on the bacterial chromosome implies that the two are similar or identical regions.

There is no direct demonstration that structural similarity plays any role in prophage attachment. Our reasons for invoking it are twofold: (1) It is known that crossing over can occur between similar (homologous) regions and (2) there is no other known basis on which two regions of nucleic acid can specifically recognize each other. Neither argument is very strong; especially since, as we shall develop later, the phage apparently does not make much use of the host recombinational system to accomplish normal integration.

Available physical information is not decisive. About 30 percent of the DNA of phage lambda is hybridizable with *E. coli* DNA. This homology is not restricted to the *b2* region, but is found in several other parts of the molecule as well. It is not known how much of the lambda *E. coli* homology involves bacterial regions close to the lambda attachment site rather than distant from it.

Whereas the physical studies indicate a more complex situation than might have been anticipated, there is nothing as yet to discourage the idea that the crossover which integrates the prophage takes place between two regions with similar base sequences.

As the *att* sites of phage and bacterium are unlikely to be identical, the following nomenclature (adapted from Signer, 1968) is designed to be nonprejudicial yet descriptive. The attachment sites for different phages are clearly distinguishable, lambda attaches at *attλ*, phage 80 at *att* 80, etc. (cf. Figs. 1-3, 4-1). Phage lambda has a region *attλ-PP* which includes the actual locus of crossover and also adjacent regions having structural effects on attachment. The attachment site of lambda on the bacterial chromosome is denoted *attλ-BB*. Prophage integration creates attachment sites that are hybrid in nature, perhaps unique for each integration event. If such a hybrid site has arisen by a crossover in which its left end derives from *E. coli* and its right end derives from lambda, the site is called *attλ-BP*. Thus in Fig. 5-1, the structure of a single lysogen would be

gal attλ-BP N R A J attλ-PB bio

This nomenclature is precise only if every integration and detachment crossover takes place at the identical point. Otherwise, any lysate prepared by inducing a lysogen will comprise a collection of phage particles nonidentical in the attachment region. The utility of this terminology depends on the fact that the crossover locus seems to be restricted to an area relatively small compared to flanking regions that exert profound structural effects on integration. Its inadequacy to handle all situations is illustrated by Six's experiments on P2 that we described in Chap. 5 where the host sites *att2-BBI* and *att2-BBII* could become incorporated into the phage particles. Whereas the terms *attλ-PP* and *attλ-BB* are here defined so as to include flanking regions such as *b2*, we will reserve the term *attλ* site to refer to the crossover point. Where a map position is indicated for *att* (as in Fig. 2-1), it is the position of the *attλ* site.

STERIC HINDRANCE AND FACILITATION

In the previous chapter, we recommended that the model of Fig. 5-1 should be adopted as a serious working hypothesis—and that we should boldly make whatever accessory assumptions are necessary to rationalize any observed facts that do not follow naturally from it. One such fact is the "steric hindrance" effect.

It has long been known that it is harder to lysogenize a lysogenic strain than a nonlysogenic one. If a lambda lysogen is superinfected with a mutant of lambda, the fraction of infected cells that become lysogenic for the superinfecting phage is less than in a control experiment where a nonlysogen is infected with the same lambda mutant. Figure 5-1 would of itself suggest the reverse. If integration is due to pairing and crossing over in regions of similar base sequence, the chromosome of the lysogenic cell offers much more similarity to the entering phage particle than does that of a nonlysogen. This should render lysogenization easier rather than more difficult.

A prophage might in principle interfere with lysogenization in one of two ways: (1) it might elaborate physiological factors that inhibit the ability of the entering particle to attach to the chromosome. We know that it makes one factor—the immunity repressor—that interferes with both function and replication of an entering particle (Chap. 9). (2) Instead of, or in addition to, such a physiological effect, there might be a structural effect of prophage on the attachment process itself.

The first experiments that distinguished these two types of effects were those of Erich Six. Six took advantage of the fact that whereas phage P2 can attach to more than one site on the bacterial chromosome, it much prefers to attach at a certain site (called position I). In a strain carrying P2 at another site (position II), the preferred location is unoccupied, but the cell is immune. This allows assessment of the effect of immunity independently of steric effects at the preferred location.

From the measured lysogenization frequencies of lysogens carrying P2 at positions I and/or II, Six concluded that both physiological and steric factors operate to reduce lysogenization frequencies. Immunity reduces the frequency of lysogenization, but the presence of a prophage at the preferred site reduces it still further. This effect was called *steric hindrance*. It constitutes a real anomaly in terms of our model.

The effect was reinvestigated by a different method in bacteriophage lambda (Campbell and Zissler, 1966). Use was again made of the partial diploid strain diagrammed in Fig. 6-2. It was then superinfected with another lambda mutant, and double lysogens were isolated. The idea was to study the purely steric effect of prophage on attachment by comparing the frequencies of addition to a vacant *vs.* an occupied site in a common cytoplasm.

The result was that the frequency of doubles was quite low, but those found were all *cis* doubles. Steric hindrance should produce an excess of *trans* doubles. The exclusive occurrence of *cis* doubles indicates, rather than a hindrance, a relative facilitation of attachment by the presence of prophage. This agrees with the simplest predictions of the model. As the observed lysogenization frequency is lower than for infection of nonlysogens, there must be a strong physiological inhibition of prophage attachment in the immune cell. This inhibition constitutes a new element added to the simple picture of Fig. 5-1.

The situation is actually even more complex, however. The observed facilitation was in immune cells, and therefore under conditions where attachment is inhibited. There might be two processes by which prophage attachment can occur: (1) a normal process, inhibited in the immune cell, where a crossover occurs specifically at $att\lambda$-BB; (2) an abnormal (or normally inconsequential) process, where a crossover occurs purely by virtue of similarity in base sequence. The second process would become quantitatively important only when the normal mechanism is repressed.

If this is true, steric facilitation should be an indirect consequence of immunity. Superinfection of a nonimmune lysogenic diploid should show no facilitation. This situation is realized by using as prophage either a heteroimmune relative of the superinfecting phage, such as lambda *imm434,* or the cryptic lambda prophage, which carries much of lambda but lacks the immunity region. In both cases, and also in mixed infection of a nonlysogenic diploid, the observed doubles are about 80 percent *trans*: 20 percent *cis.*

This shows that, in the normal addition process, there is not a facilitation but a hindrance. Assuming the existence of a specific catalyst for crossing over at the insertion site, the simplest expectation would be ⅔ *cis*: ⅓ *trans*. The observed results are far from that, and therefore some addition al hypotheses are necessary to explain them completely.

If the effect of immunity on prophage attachment is to repress synthesis of a diffusible "attachment enzyme," this enzyme should be able to function for any phage genome in the cell, even one whose ability to make the enzyme is blocked by immunity. This point was tested by simultaneous superinfection with a homoimmune and a heteroimmune phage. By itself, the homoimmune phage would always add *cis*; the heteroimmune one would add either *cis* or *trans*. However, in mixed infection, the presence of the heteroimmune phage allows the homoimmune phage to also add *trans* as well.

Teleonomically speaking, the results are about what might be expected. Assuming that lysogeny has a survival value for temperate phages, it is reasonable that attachment should take place with high frequency when the phage wants it to happen (after infection) and that, after lysogeny is established, the system should be inhibited so that it will not detach the phage from the chromosome except in rare cells where immunity breaks down and phage growth can occur. Elaboration of a specific repressible recombinase that recognizes the *att* region is a reasonable answer to this problem. The observed steric hindrance in nonimmune cells indicates that the "natural" mode of addition (recombination between $att\lambda$-PP and $att\lambda$-BB) is favored above any unnatural mode, including crossover at the termini of an integrated prophage ($att\lambda$-$PP \times att\lambda$-PB or $att\lambda$-$PP \times att\lambda$-BP). This finding in any case indicates that $att\lambda$-PP and $att\lambda$-BB are not equivalent.

The above explanation implies that, in homoimmune infections where addition is always *cis,* the second phage should frequently be integrated interstitially within the first one, whereas in heteroimmune or nonimmune hosts, integration should always be at the *att* site. Both expectations seem to be fulfilled (see Chap. 12).

Thus in phage lambda, either hindrance or facilitation can be observed, depending on conditions. In phage P2, only hindrance has been found. Results with these two phages are not entirely comparable in that *cis* double lysogens are never formed by phage P2. When P2 adds to a strain that is already lysogenic, it either replaces the preestablished prophage or attaches at some previously unoccupied site.

The physiological repression of prophage addition seems to be much stronger for lambda than for P2. The effect of immunity on lysogenization by P2 can be explained as an indirect consequence of the inhibition of phage multiplication (Six, 1961). This cannot be the case for lambda, where immunity reduces prophage attachment even for early *sus* mutants that do not multiply in any event. There must therefore be a strong direct effect of immunity on some step of lambda prophage integration.

From experiments on prophage curing (described in Chap. 7), it seems that different phages such as lambda and 80 elaborate different catalysts for integration, each specific to its own site. It is therefore possible to create in lambda the equivalent of the P2 system. To a hybrid phage with the immunity of lambda and the insertional specificity of phage 80, an ordinary lambda lysogen constitutes an immune cell in which the "preferred site" is unoccupied. A strong inhibition of prophage attachment is observed in this situation (Taylor and Yanofsky, 1966).

Experimental results on steric hindrance and related questions, employing partial diploid bacterial strains, are summarized in Table 6-1. From these experiments, together with those of Signer and Beckwith (1966) and Fischer-Fantuzzi (1967), the following tentative conclusions have been drawn: (1) Lambda and related phages synthesize a catalyst, presumably an enzyme, causing prophage attachment. This catalyst we call "integrase." (2) Integrase is site-specific. (3) Integrase production is repressed in the immune cell. (4) Integration can occur at low frequency in the absence of integrase. Such integration is dependent on the bacterial recombinase, and is facilitated by increasing the similarity in base sequence between the attaching phage and the bacterial chromosome. This is the explanation of steric facilitation. (5) When attachment is mediated by integrase, the presence of prophage creates a steric hindrance. The reason for the hindrance is unknown.

TABLE 6-1. *Lysogenization of Partial Diploid Bacteria*[a]

Infecting Phage	Cell	Addition	Type of Addition			Ratio
			cis	trans	Other[b]	(cis: trans)
λ + λ	—	λ + λ	22	66	8	22:66
λ	λ	λ	64	1	0	64:1
λ	λ	λ	12	32	3	15:35[c]
434	434	434	33	0	0	33:0
λ	434	λ	3	13	0	3:13
λ	λcry	λ	5	35	8	13:43
λ + 434	434	λ	19	29	2	21:31
λ + 434	434	434	30	22	0	30:22
λ + λb2	—	λ + λb2	23	0	7[d]	23:0
λint + 434	434	λint	30	0	0	30:0
λint + 434	434	434	64	5	0	64:5

[a] Summary of results with partial diploids. Taken from Campbell (1965a,b), Campbell and Zissler (1966), Zissler (1967), Zissler and Campbell (1968). Types of addition classified as in Fig. 6-2. λ and 434 refer to immunity type. The 434 hybrid phage lambda *imm434* was used.

[b] Triple or quadruple lysogens, not completely characterized in all cases. Where both *cis* and *trans* additions in the same cell could be classified, these are included in the totals of the final column.

[c] Aberrant results obtained with one lysate during one time period, for unknown reasons.

[d] In all these 7, lambda *b2* was *cis* to another prophage.

GENETIC FACTORS IN PROPHAGE INTEGRATION

If integrase is a phage-specific protein, phage mutants should occur in which the quality or quantity of integrase is altered. Such mutants are known both in phage P22 and phage lambda (Smith and Levine, 1967; Zissler, 1967; Gingery and Echols, 1967; Gottesman and Yarmolinsky, 1968). The P22 mutants are called *L* (for lysogeny) and the lambda mutants *int* (for integrase).

The *int* gene maps between *b2* and *cIII*. The mixed superinfection experiments shown in Table 6-1 indicate that *int* mutants are defective in production of a diffusible attachment factor. Lambda *int* lacks the ability, which lambda *int*$^+$ has, of allowing lambda *imm434* to attach at an unoccupied site in an immune cell. This experiment also suggests that the *int* gene is controlled directly by the immunity; the *int*$^+$ gene of the *imm434* phage fails to function, even though lambda is present to provide any earlier phage functions necessary to induce this enzyme. Like the P22 *L* mutants (and unlike lambda *b2*), lambda *int* can form single lysogens in mixed infection with wild type. A second gene (*intB*) necessary for stable integration of lambda has been described by Gingery and Echols (1967). It maps between genes *P* and *Q*. Its function is unknown.

As mentioned earlier, integrase (defined physiologically) is site specific. The *int* gene product also carries such site specificity. Whereas wild-type lambda will help lambda *int* to integrate, a hybrid phage with the attachment specificity of phage 21 fails to do so.

Genetic recombination in lambda can be caused by at least three determinants—the recombinase (*rec*) genes of the host, a gene (*red*) of the phage, and *int*. In the absence of all three, recombination is nil. Whereas *rec* and *red* cause recombination throughout the phage chromosome, *int*-mediated recombination is almost exclusively in the region between *b2* and *cIII*, presumably at the *att* site (Echols, Gingery, and Moore, 1968).

If lambda produces a specific recombinase catalyzing its own insertion, then prophage insertion might take place in *rec*$^-$ bacteria. It does (Brooks and Clark, 1967). On the other hand, insertion of a defective phage that lacks integrase activity should depend on the *rec* gene. This is true for the defective transducing phage 80 *dlac*, which integrates in the *lac* region of the bacterial chromosome under these circumstances (Signer and Beckwith, 1966).

If our interpretation is correct, lambda integrase is the first example of a specific recombinase. The ability of a protein to recognize a specific nucleotide sequence and interact with it is not a new idea, however: Repressors, amino acid activating enzymes, and methylating enzymes must be similarly endowed. The biological behavior of a DNA sequence derives equally from its intrinsic properties and from the various recognition systems in the cell that contains it. Some of these, such as those directly involved in coding, are common to all cells and recognize configurations that frequently recur in every large DNA molecule. The systems mentioned above are specific to particular cells and must recognize rare configurations, perhaps sequences sufficiently long to be unique.

Since the recombinase is specific, we must be careful not to assume that prophage insertion resembles normal recombination in any properties not directly tested on the integrase system itself. For example, there is no necessity to assume that the two partners in the recombination (*att-BB* and *att-PP*) contain similar base sequences. The fact that integrase can mediate recombination between two *att-PP* sites encourages the belief that they do. Homologous recombination remains the simplest possibility.

Other episomes may differ from lambda in important details. In phage P22, for example, integrase synthesis is not completely repressed by immunity. Temperature shift experiments with an integrase mutant show that integration is augmented by restoration of the permissive temperature long after immunity is established (Smith and Levine, 1967). The very low rate of spontaneous integration of the F factor suggests that this episome does not form such a system, or else that it is repressed under most circumstances. Occasional situations where F comes out of the chromosome at high frequency are intriguing with respect to the latter possibility (see Chap. 7).

Given the existence of a site-specific integrase, we can examine the meaning of the structural specificity of the *att-PP* and *att-BB* regions. One or both of these regions should contain a site of integrase action, as well as one or more recognition sites for the enzyme. None of these sites need be identical with the "att site," which we have defined as the crossover point. It is possible, but not mandatory, that the two regions *att-PP* and *att-BB* should interact with each other, as by base pairing of similar sequences. Such interaction sites need not be identical to any of the sites recognized by integrase. In later chapters we will use the term *att* to refer sometimes to the integrase substrate and sometimes to the crossover point. This is because we are discussing a complex genetic region whose finer structure is not yet clearly resolved.

7

MECHANISM OF DETACHMENT

In lysogenization, the phage genome is inserted into the bacterial chromosome. When a lysogenic bacterium makes phage, either spontaneously or following induction, the phage genome must be excised from the bacterial chromosome. Physical excision has been demonstrated under certain circumstances (Ptashne, 1965a). It is not yet proven that this is a major route by which phages normally arise from lysogenic cells; but, as in the previous chapter, we shall assume that it is. Lysogenic bacteria can also give rise to nonlysogenic "cured" progeny. Physical excision is assumed (but not demonstrated) in this case as well.

According to Fig. 5-1, the mechanism of prophage insertion is crossing over between *att-PP* and *att-BB*. The mechanism of detachment would be the reverse.

It is a well-known thermodynamic principle that a catalyst will accelerate forward and reverse reactions equally. Care is required in applying this principle to the phage integrase, because the diagram of Fig. 5-1 represents a genetic scheme, not a balanced chemical reaction. The actual reaction might be a simple exchange as drawn, but could as easily be an essentially irreversible process, coupled for instance to the hydrolysis of adenosine triphosphate (ATP). The catalyst would of course catalyze the reverse chemical reaction (with concomitant condensation of ADP and iP to form ATP), but this would not constitute a significant pathway for excision.

If *att-PP* and *att-BB* were physically identical, as implied by Fig. 5-1, the attachment reaction would be symmetrical with respect to the chromosomal regions actually participating. Attachment and detachment would then be chemically the same reaction, and the same catalyst should mediate both, whether or not the relevant chemical process is reversible. We have indicated in the previous chapter that the two sites are not completely equivalent, but that some similarity between the two is likely.

The available facts are limited. Single lysogens of P22 mutants, which are blocked in prophage attachment, fail to produce much normal phage after induction. Abnormal, defective particles are formed, indicating abnormal detachment, or perhaps replication without detachment (H. Smith, 1968). Lambda *int* mutants behave similarly. Both lysogenization and heteroimmune curing (see below) are reduced at temperatures above 40°C (Campbell and Killen, 1967). It seems likely that insertion and excision have some steps in common, but we cannot say whether a precise reversal is involved.

CURING

Two methods have commonly been used for curing lysogenic bacteria. The first is to expose the cells to high doses of irradiation. The second is to infect them with a weakly virulent or heteroimmune relative of the prophage.

Radiation Curing. Nothing is known about the mechanism of radiation curing. Among ninety-six isolates cured of phage lambda immunity by X-irradiation, none carried any cryptic phage genes detectable by marker rescue, and none was missing the nearby chromosomal genes for galactose metabolism (del Campillo Campbell, 1968). What is deleted in curing is precisely what was added at lysogenization —no more and no less.

Superinfection Curing. Superinfection curing has been studied more intensively. Presumably the integrase of the superinfecting phage detaches the resident prophage. Integrase mutants cause little curing when they superinfect heteroimmune lysogens. Curing is site specific. Only those heteroimmune phages attaching at the same chromosomal site as the prophage will cure. Thus, the lambda-434 hybrid phage, for instance, will cure lambda. Actual attachment by the superinfecting phage apparently is unnecessary. The *b2* mutant of lambda, for example, will cure 434 lysogens (Signer, 1968).

The general picture, as outlined already in the previous chapter, is that each phage carries a gene for its own kind of integrase, and an *att* region which, together with the corresponding *att* region of the host, serves as a substrate for the enzyme. When entering a nonimmune cell, the phage causes synthesis of its specific integrase. If the cell contains a prophage whose termini constitute a substrate for that integrase, it stands some chance of being excised and discarded.

Spontaneous Prophage Loss. With most lysogenic systems, the probability of spontaneous prophage loss during growth is low but not zero. Accurate measurement requires prevention of reinfection within the culture. Prophage loss should entail a reversal of insertion —pairing between the ends of the prophage, ejection of a circular phage chromosome, and restoration of the nonlysogenized chromosome. The fact that the rate is low reinforces the conclusion of the previous chapter that the amount of base sequence similarity is limited at best, so that recombination in this region becomes appreciable only following induction of a specific integrase.

The extent of base sequence similarity can be increased by use of a transducing phage such as lambda *dg* (see Chap. 8). In this case, the prophage is surrounded by an extensive tandem duplication, recombination in which will cause prophage loss. The rate of loss should therefore be higher than from a normal lambda lysogen, and in fact it is.

Induction Curing. If a lambda mutant that makes a temperature-sensitive repressor is heated, phage production is induced. A short temperature exposure, insufficient to induce phage production, can result in a high survival of cured cells (Weisberg and Gallant, 1966). Curing can also occur when the complete induction process is arrested by a mutational block rather than by transient heating. All of these mutants are defectives or conditional lethals with early blocks in DNA synthesis, but not all early mutants cause curing. In lambda, extensive curing is seen with mutants in the X region, and less so with O and P mutants. N mutants are not cured. When curing occurs, survival is higher than when wild-type lambda is induced; so some cells that would have died and produced phage in a wild-type lysogen are instead cured in the presence of the mutant block.

Thus there seems to be an intermediate stage in normal induction, at which interruption of the process leads to curing. The actual time of prophage detachment cannot be inferred from these observations; only the time of commitment to detach. It is possible that radiation curing at high doses bears some relation to induction curing.

DETACHMENT OF THE F FACTOR

The change from the F^+ to Hfr state is reversible. There is general agreement that the rate of reversion from Hfr to F^+ varies widely from one Hfr to another; but rate measurements are hard to find.

The behavior of attached F during P1 transduction suggests that F detachment may sometimes occur at a much higher rate than normal.

When F itself is transduced from an Hfr donor into an F^- recipient, it is usually recovered in the autonomous rather than the integrated form. When an Hfr strain is used as a recipient in transduction for a chromosomal marker closely linked to the site of F attachment, many of the transductants are F^+ rather than Hfr (Pittard, 1965).

8

ABNORMAL DETACHMENT AND THE FORMATION OF TRANSDUCING PHAGES

GENE PICKUP BY EPISOMES

When a lysogenic bacterium produces phage, either spontaneously or following induction, the lysate may contain (besides ordinary phage particles) rare exceptional particles that include some bacterial genes. These particles are detected by their ability to transfer these genes to another bacterium of appropriate genotype. In the case of phage lambda and its relatives, the only bacterial genes transferred are those very close to the prophage on the bacterial chromosome. Those particles harboring the bacterial genes also include some or all of the phage genome, and the phage genes and bacterial genes in this particle are connected on one DNA molecule. Furthermore, such particles arise only when the phage comes from a lysogenic bacterium, not during lytic multiplication.

One of the principal motivations for the model of Fig. 5-1 was that it provided a simple explanation for the origin and properties of the transducing phage lambda *dg*. At the time, the data on lambda *dg* were more complete and convincing than those on prophage mapping.

Lambda *dg* (defective, galactose-transducing) is the name given the lambda phage variant that includes galactose genes of the host. It turns out that such variants are always defective phages containing some but not all of the lambda genome.

How do such variants originate? We cannot provide a detailed mechanism, but can circumscribe the question to some extent: Because transducing variants arise only in lysogenic bacteria and include only genes close to the prophage, we suppose that they preserve the same connectedness that existed in the lysogenic chromosome. Just as each deletion mutant is missing a connected block of genetic material, each transducing phage should comprise such a connected block. A genetic map of the lysogenic chromosome can then be constructed from the family of transducing phages, just as it can be from a family of overlapping deletions.

When this is done with lambda *dg*, the resulting map is precisely that deduced in Chap. 5 from the results of bacterial crosses, i.e., the gene order is permuted as shown in Fig. 5-1. The same conclusion has since been verified for the lambda variants transducing *bio*. Transducing phages that include some but not all of the *gal* or *bio* regions can be used to determine the gene order within these regions and also their orientation with respect to the prophage. This method agrees with other mapping procedures and is probably more reliable than crossing data in determining gene order.

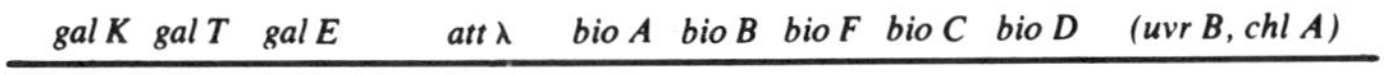

| galK galT galE | attλ | bioA bioB bioF bioC bioD | (uvrB, chlA) |

FIG. 8-1. Order of genetic loci in the vicinity of *att*λ. *galK, galT*, and *galE* are the structural genes for galactokinase, galactose-l-phosphate uridyl transferase, and uridine diphosphogalactose-4-epimerase, respectively. They constitute an operon transcribed from right to left on the map as drawn. The five *bio* genes are concerned with different steps of biotin biosynthesis, which are not yet known in complete detail (del Campillo *et al.*, 1967; Rolfe and Cleary, 1968). The *uvrB* gene (Howard-Flanders *et al.*, 1966) imparts UV resistance to wild-type *E. coli*, probably by synthesis of a repair enzyme. Mutants or deletions of this gene are abnormally sensitive to ultraviolet light. *chlA* is one of the genes of the nitrate reductase system. Nitrate reductase can also reduce chlorate to chlorite, which is toxic to *E. coli*. Under anaerobic conditions, where nitrate reductase is induced, wild-type *E. coli* dies on chlorate medium. Mutants in which *chlA* is deleted or nonfunctional are chlorate resistant.

The genetic map of the *E. coli* chromosome in the neighborhood of *att*λ is shown in Fig. 8-1. The results of topological mapping of this region, by use of bacterial deletions and transducing phages, are presented in Figs. 8-2 and 8-3. To ensure valid comparison, only data collected in the author's laboratory are included in these figures.

My ideas on the origin of lambda *dg* are diagrammed in Fig. 8-4. Two features should be stressed: (1) I suppose that, as in the origin of deletion mutants, rare breakage and joining leads to the union of essentially heterologous genetic material. I am not suggesting that the formation of lambda *dg* from a lambda lysogen depends on homologous pairing and crossing over as the formation of lambda itself does. We are speaking of a rare, exceptional event that happens about once in every 10^5 phage particles. (2) Such heterologous recombination is

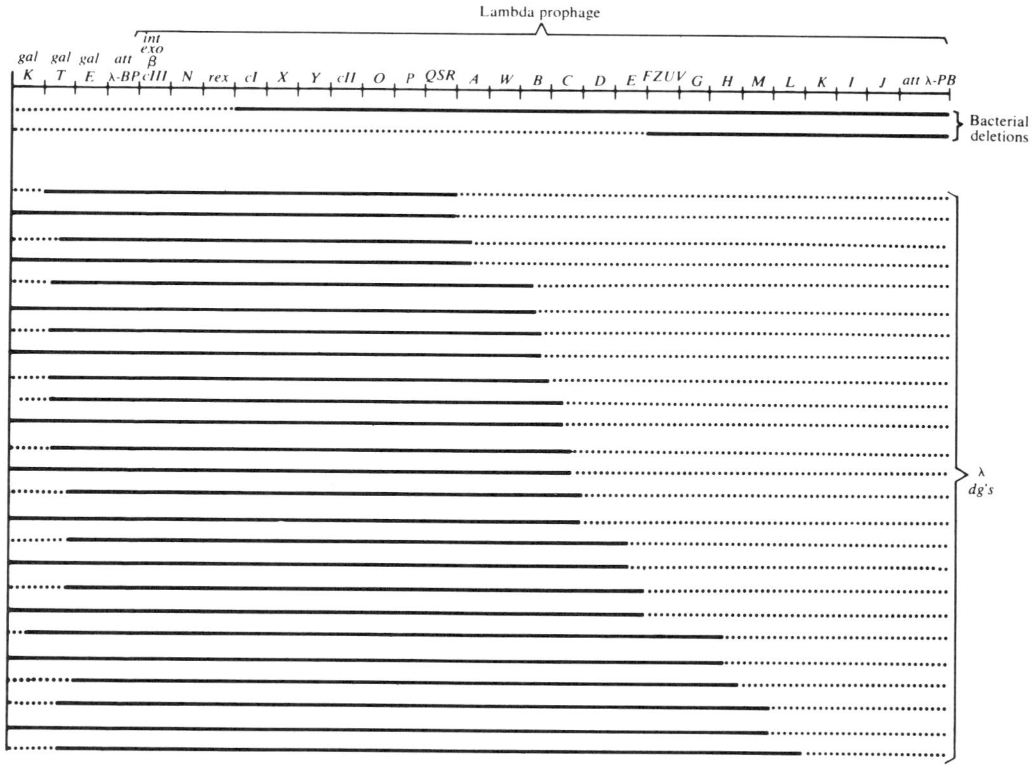

FIG. 8-2. Topological mapping of lamba prophage and genes to the left. Bacterial deletion mutants were isolated for chlorate resistance by Adhya (1968). The chlorate resistance of these mutants indicates a *chl* locus (as yet unidentified) between *gal* and lambda. A larger collection of deletions at this end of the lambda prophage was assembled by Shapiro (1967), using a different method of isolation. Lambda *dg*'s are from Kayajanian and Campbell (1966). Solid lines indicate regions present. Dotted lines indicate regions absent. Vertical lines on top indicate cistron boundaries.

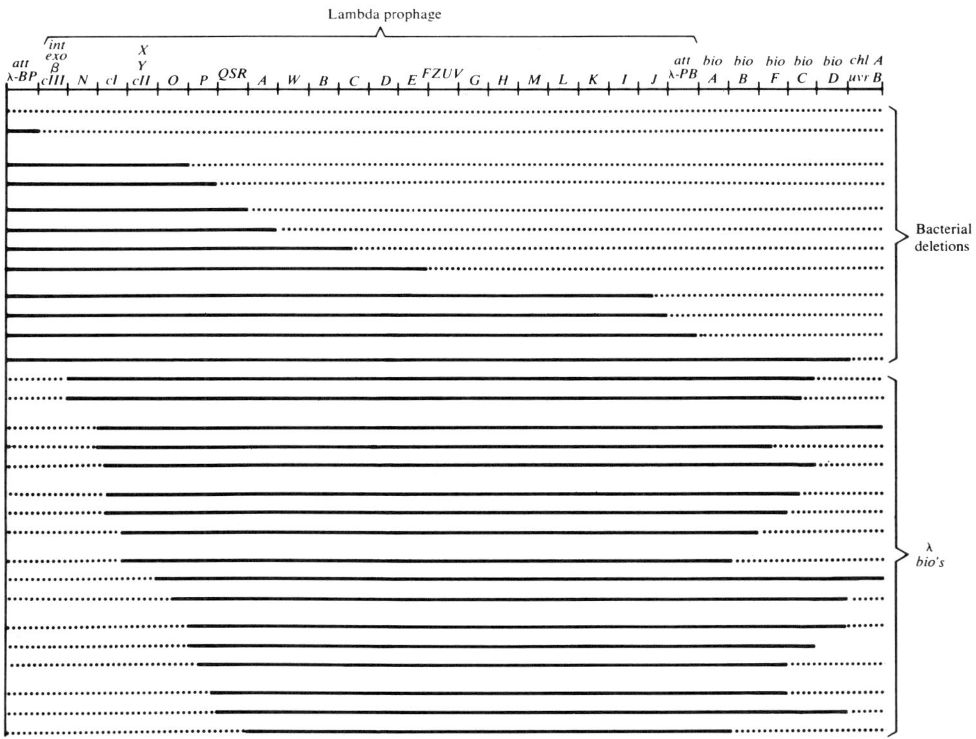

FIG. 8-3. Topological mapping of lambda prophage and genes to the right. Bacterial deletions were isolated for chlorate resistance by Adhya (1968). Lambda *bio*'s were isolated by Kayajanian (1968).

GENE PICKUP BY EPISOMES   103

required only for the origin of lambda *dg*, not for its subsequent reproduction. Once formed, the lambda *dg* can reproduce either as an autonomous phage or as a prophage in the same way that lambda itself does. No additional unlikely events are required. This agrees with the observed fact that the properties of lambda *dg*, both its gene content and its density, are stable hereditary properties that do not change once the lambda *dg* is formed.

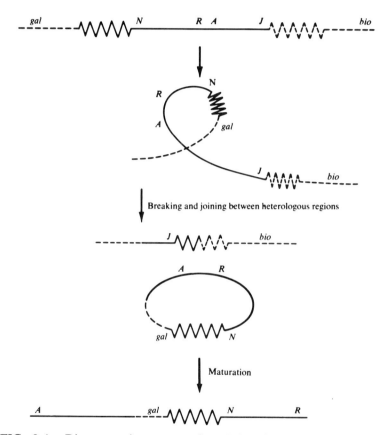

FIG. 8-4. Diagrammatic representation of the origin of lambda *dg*.

Lambda *dg* thus bears the same relationship to lambda as does a chromosomal aberration of a higher organism to its normal counterpart. It is formed by a rare unique event that breaks and joins DNA in regions having little or no similarity in base sequence. Once formed, it is quite stable. Mutant alleles of the genes it contains can be introduced into it by normal recombination without changing the basic character of the aberration itself.

As might be expected if heterologous breaks are involved, there is no predictable relationship between the locations of the two break points of the same lambda *dg*. Both the *gal* region and the phage genome are divisible into many subregions. Different lambda *dg*'s terminating in the same region of lambda may extend into different subregions of *gal* or may contain the entire *gal* operon. When the location of both break points is known, physical measurements show complete agreement between the amount of DNA present (as reflected in the density) and the amount of genetic material present according to the postulated mode of origin (Kayajanian and Campbell, 1966).

It is possible that heterologous breaking and joining is reciprocal. In that case, many deletion mutations would be accompanied by formation of a fragment whose ends would join to form a ring. Usually, such a fragment would be lost because it lacks an initiator for replication. However, a fragment containing enough phage genes should be able to multiply, mature, and lysogenize like a phage and therefore should persist genetically.

How many genes are enough? Some transducing variants (such as certain isolates of lambda *bio* or phage 80 *trp*)[1] contain enough of the phage genome to make plaques. Others, such as lambda *dg*, are missing phage genes needed for lytic growth. They are defective. Their ability to mature into complete particles depends on the presence in the same cell of a normal phage containing the genes they lack. The lambda *dg*'s most deficient in phage genes are those that lack the entire region *A-J*. This fact suggests that the ability to go through a complete infectious cycle, even in the presence of a normal phage, requires that the ends of the vegetative chromosome be present.

Starting from the other side, we can look at biotin transduction by lambda. Some transducing variants may contain a complete lambda genome, but others lack a portion of it. As *bio* is on the opposite side of lambda from *gal*, the gene content of defective biotin transducers is different from that of lambda *dg*: Lambda *db*'s always have genes *A-J*, but can be missing *N*, or in the most extreme case, all the genes from *N* to *R*. We conclude that the only parts of the lambda genome that can be *structurally* essential for a complete lysogenic cycle in the presence of helper phage are the vegetative ends and the attachment locus *att*. The other genes of the phage are necessary only for physiological reasons; their products can be supplied in *trans*. These facts are of some interest with respect to the location of the initiation site for DNA synthesis in lambda (cf. Chap. 10).

[1] Phage 80 will transduce the adjacent gene cluster governing tryptophan synthesis just as lambda will transduce the biotin cluster.

We may speculate that lambda *dg* and a bacterial deletion mutant might be products of the same event. Such speculations must be clearly distinguished from demonstrated facts. This point cannot be tested directly because the cell from which lambda *dg* derives must lyse in order to liberate the transducing particle. There is no evidence that lambda *dg* can arise in a cell before that cell is committed to lysis.

The detailed mechanism of origin of transducing phages, as of chromosomal aberrations in general, is still shrouded in mystery. The process does not seem to require participation of the phage-specific system involved in normal excision; yet the fact that transduction by lambda is restricted to markers in the vicinity of the prophage suggests that something other than random fragmentation of the entire bacterial chromosome is involved.

The F episome resembles phage lambda in occasional occurrence of variants that have picked up genes from the bacterial chromosome. As with lambda, such variants are produced only from integrated F (i.e., from Hfr strains) and seem to comprise continuous blocks of genes contiguous to the F attachment site of the particular Hfr strain used.

The presence of such modified F factors (F') is detected by their ability to transfer at high frequency those bacterial genes associated with F. As the F' can multiply indefinitely in the bacterial clone of origin (without killing the cell, as lambda does), it is possible to examine whether F' formation is associated with a concomitant deletion of genetic material from the bacterial chromosome. At least for some F' factors, it is (Scaife and Pekhov, 1964; Berg and Curtiss, 1967). The event may not be exactly reciprocal. As mentioned in Chap. 5, integration of F sometimes seems to cause deletions or disruptions of genes in the neighborhood of the integration site. This could likewise be true of the excision events.

Given that episomes "pick up" chromosomal genes as the result of rare crossover events between heterologous regions, there are obviously two basically different types of abnormal excision possible— which have been called Type I and Type II (Scaife, 1967). Figure 8-5 shows the essential difference. In a Type I event, one break takes place within the integrated episome and the second within the bacterial chromosome. A Type II event, on the contrary, involves two breaks within the bacterial chromosome, one on either side of the integrated episome.

The known transducing variants of phage lambda are all of Type I. This is explainable by the fact that the variant chromosome, to be observed, must be of a size that will fit into the phage head. This probably limits the possible amount of excess DNA to about 10 percent. It may also be that a Type II particle, if formed, would become dissociated some time during its life cycle into a normal phage and a bacterial fragment, due to integrase action.

In the case of F, no known factors limit the size of the autonomous element, and no specific system for excising the integrated episome is known. The products of Type II events should be more readily recoverable than they are with lambda, and such events do in

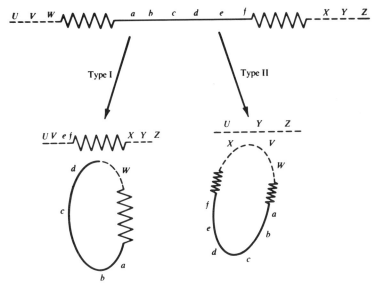

FIG. 8-5. Diagrammatic representation of two alternative types of abnormal excision of episomes. *UVWXYZ* are bacterial genes; *abcdef* are episomal genes. In Type I, bacterial and episomal genes become fused by heterologous breaking and joining—as shown for the specific case of lambda *dg* in Fig. 8-4. In Type II excision, the excision event directly involves only bacterial regions. The immediate neighborhood of the episome is unaffected.

fact occur: F' factors carrying markers both distal and proximal to the site of F integration are thus formed. It is in these cases that the reciprocal "crossover product" has been demonstrated.

The point to note about Type II events is that they do not directly involve the episome at all. They are basically crossovers between heterologous parts of the bacterial chromosome, which reduce the circular chromosome into two smaller circles. Either circle might in principle carry the episome.

What distinguishes the formation of an F' element on the one hand, from a minute chromosomal deletion in some region far from the site of F integration? Topologically, the two events are identical. The large circle has been reduced to two smaller ones, one of which carries F. Is the difference merely a matter of size, or is there some factor of qualitative interest not included in the diagram of Fig. 8-5?

If the bacterial chromosome has a unique site for initiation of replication, the outcome should be determined by whether or not the crossover event separates the episome from that bacterial initiation site. If the two are separated, one unit will behave as the chromosome and the other as F'. If the two are together, the other product will have no site of replication origin and will be lost from the progeny through failure to replicate.

The relative size of the two products would then not be of primary interest—except insofar as loss of large sections of chromosomes might prove lethal. It should even be possible for an Hfr strain to decompose into an F' containing almost all of the bacterial chromosome, accompanied by a small block of nonessential genes connected with the normal bacterial replicator region. We might well inquire "Who is the episome?" in such a case. Effectively, the F factor would have taken over the normal replicator function for the major part of the bacterial chromosome.

As Type II events do not disturb the relationship of the episome to its neighbors, it should be possible for normal F to dissociate from F'. This situation has been reported.

Presumably both F and lambda can undergo Type I and Type II events. Type II events are seldom if ever recoverable in the case of lambda. It is less certain whether any known F' results from a Type I event. Defective mutants of F are known. None has been found to arise in concert with acquisition of bacterial genes; but it is likely that defectives of the known types would not be found in the ordinary selection procedures for F', even if they occur. The bacterial product of a Type I event would carry part of the sex factor still integrated into the chromosome. Presumptive strains of this type are found. They have an increased affinity for normal F at the original site of integration. We do not know, however, whether Hfr strains can carry more than one copy of F in tandem, like prophages of lysogenic strains. The F component of known F' strains may always be complete and potentially dissociable from the F' complex.

Certainly our knowledge of Type I events is too meager for us to generalize the Type II result and say that Type I events also create a bacterial product exactly reciprocal to the derived episome. It is hard to see why the same mechanism operative for Type II events should not be responsible for at least a fraction of the Type I events, however.

GENERALIZED TRANSDUCTION

Having made some headway in understanding specialized transduction, we can consider generalized transduction in perspective. Phages P22 and P1, unlike lambda, can carry any known bacterial gene from one cell to another and can pick up genes in a cycle of lytic growth, without going through the prophage state.

Because transductants frequently were not lysogenic, Zinder and Lederberg originally proposed that transducing particles contain loose fragments of the bacterial genome, produced by nuclear breakdown of the host. These could occasionally become wrapped up in the coat of a phage particle in addition to (or instead of) the phage chromosome. After the results of lambda transduction became clear, it was possible to imagine that, instead, transducing particles contained a "hybrid" DNA molecule in which phage genes and bacterial genes are attached in linear sequence. Depending on how defective the phage genome was, such a molecule might be able to form plaques, to lysogenize without forming plaques, or only to pair with its homologue in the recipient cell and exchange genes with it. This viewpoint was bolstered by the discovery of Luria's group that phage P1, which ordinarily shows generalized transduction, can be made to exhibit specialized transduction as well (Luria et al., 1960).

If phage P1 grown on lac^+ *Escherichia coli* is added to lac^- cells, lac^+ transductants appear. These are stable lac^+ which show no evidence of carrying anything analogous to lambda *dg*. However, if the same P1 lysate is applied to bacteria of the genus *Shigella* which are permanently lac^- and lack anything similar to the *E. coli lac* region, pairing and exchange with the chromosome cannot occur. Under these circumstances, persistently unstable lac^+ transductants are found, which harbor a genetic element containing some P1 genes and the *lac* region of the bacterium. From these bacteria, with the help of a good P1, high frequency transducing lysates active on *E. coli* as well as *Shigella* can be obtained.

This gave us a clear prototype for explaining all generalized transduction: There is always a "hybrid intermediate" structure, but it can be elusive. By being especially clever and putting it into *Shigella*, Luria had trapped it so we could see it.

GENERALIZED TRANSDUCTION   109

Subsequent work, while casting no doubt on the validity of Luria's observations or of his explanation for them, has shown that such hybrid particles do not play a major role in generalized transduction. Instead, most transducing particles appear to contain only bacterial DNA. A variety of interesting experiments on this question (e.g., Okubo et al., 1963) culminated in the definitive studies of Ikeda and Tomizawa (1965).

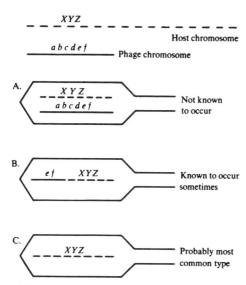

FIG. 8-6. Models for the nature of the particles responsible for generalized transduction. In (A), a small fragment of bacterial genome is included within an otherwise normal phage particle. In (B), the phage head contains one DNA molecule with phage and bacterial genes covalently linked. In the example drawn, the phage genome is incomplete (defective). In (C), a phage-sized fragment of host DNA is included in the phage head without any associated phage material.

Infection of a bacterial cell by the P1 stock they used stops host DNA synthesis completely—including that host DNA which eventually becomes part of the transducing particles. If this DNA has been labeled with 5-bromouracil,[2] the transducing activity is all located in a dense fraction of the lysate, physically separable from the plaque-forming particles. Radioactive phosphorus added after infection is

[2] DNA containing 5-bromouracil in place of the normal base thymine (5 methyluracil) has a higher density than natural DNA because the bromine atom is heavier than the methyl group.

incorporated into progeny phage, but not into transducing particles. If any phage DNA is in the transducing particles, it is below the level of detection by this method. The only possible limitation to this argument might be if transducing particles arise in a special fraction of the cell population where nuclear breakdown is extensive and therefore new phage DNA is made from preassimilated rather than postassimilated material.

The picture that emerges from Tomizawa's work is that phage P1 probably resembles the T even phages in maturing with an exact "headful" of DNA. Bacterial DNA may also be chopped to size in place of the viral DNA and wrapped up in a phage head.

Progress since the time of Zinder and Lederberg is diagrammed in Fig. 8-6. The possibility that seems definitely excluded is that shown in Fig. 8-6A. Whenever careful precautions are taken to avoid reinfection on the isolation plates, transductants prove to be almost entirely nonlysogenic. There is no reason to suppose that more than one nucleic acid molecule ever gets wrapped in the same coat.

THE CRYPTIC PROPHAGE

Phages can be defective because of the absence of genes necessary for lytic growth or because of mutation in those genes. The first defective phages observed turned up in the prophage state. They were recognized as derivatives of normal lysogens that no longer liberated phage but retained the specific immunity against superinfection by the prophage type. Induction of defective lysogens and superinfection with mutant phages showed that markers of the defective phage could be rescued by recombination into active phage.

Most defective isolates have been induced by irradiation of lysogenic cultures. This treatment leads to curing as well as mutagenesis. Nonimmune isolates have occasionally been examined also for marker rescue, but for a long time only immune defectives and completely cured strains were found.

In 1964, Fischer-Fantuzzi and Calef reported the first case of a nonimmune defective, which they termed the "cryptic prophage." This phage had suffered a deletion of phage genes including genes N through R. The left end of the vegetative map, from A to J, was present, as determined by marker rescue.

The first cryptic prophage turned up somewhat by accident, in a K-12 strain that had been "cured" of lambda several years earlier. After its discovery, Fischer-Fantuzzi and Calef went back to the original lysogenic stock from which it had been derived. When additional cured derivatives were obtained from this stock, some of them were likewise cryptic. Furthermore, if the phage from the original stock was used to lysogenize another bacterium, the new lysogen also gave rise to cryptic prophages. Lysogens of standard type lambda seldom produce cryptics; when they are irradiated, the nonimmune survivors have generally lost all trace of the phage. With the lambda stock of Calef and Fischer-Fantuzzi, however, about 50 percent of them carry cryptic prophages. Clearly there is some salient genetic difference between their phage (which is more closely related to the original lambda indigenous to *E. coli* K-12) and standard type lambda. This phage, which can generate cryptic prophages by ultraviolet curing, was called *cryptogenic* (*crg*).

In addition to the common type of cryptic prophage first discovered, two other kinds have been found, but much less frequently: (1) a cryptic prophage containing genes *A-J* and also *Q-R*; (2) one containing only genes *G-J*.

Standard type lambda can also generate biotin transducers which have the formal properties of a cryptic (Fig. 8-3). They contain some phage genes, as judged by marker rescue, but lack the immunity region. These do not fall into the same genotypic classes as do the products of curing the cryptogenic strain. All contain genes *A-J*, but the region *cI* through *R* is dissected at numerous points. Moreover, there seems to be a fundamental difference in mode of attachment. When strains carrying lambda *db* are superinfected, some lambda *db* is produced in the yield, along with phage of the superinfecting type. The cryptic prophages, by contrast, do not mature in detectable numbers following superinfection. Some of their genes find their way into mature particles, but these are active phage particles formed by recombination with the superinfecting type.

An interesting feature of the "cryptic" deletion is that, when it occurs, the linkage between the *gal* and *bio* markers that bracket the prophage is strongly reduced. This has been shown both in P1 transduction and in bacterial crosses. It is as though the bacterial chromosomes had been split apart at that point and had either remained open or (more likely) had undergone some rearrangement that introduced additional genetic material in this region. This disruption of linkage is not observed, however, in the rare cryptic type that contains *Q* and *R* and thus preserves the linkage AR across the vegetative "ends."

As to the difference between standard lambda and lambda *crg*, little is known. Even the genetic mapping of the *crg* character is not certain at present writing. The determination seems to be multifactorial, with some of the relevant determinants between *J* and *cI* (Marchelli et al., 1968). No very specific models for cryptogenicity have been proposed, but the available vague ideas can be arranged into two categories:

1. *Structural explanations.* These would say that lambda *crg* has some chromosomal aberration (inversion, translocation, duplication, etc.), such that it can give rise, by internal recombination, to lambda *cry.* It might, for example, have incorporated some bacterial genes normally distant from lambda and be able to recombine with them. It is hard to design a single aberration that would generate the rare types as well as the common ones; but the rare type, being rare, might arise by a different mechanism.

2. *Physiological explanations.* The cryptogenic phage might differ from standard type in the quality or quantity of the enzymes concerned with prophage excision and/or end joining and splitting.

In other words, cryptogenicity could be the first concrete example of a possibility we will discuss in more detail in Chap. 11. If joining and splitting of the phage chromosome betweeen R and A is part of the normal life cycle of the phage, then lysogenization creates for the host not only the hazard of full-fledged induction with concomitant phage synthesis, but also that of having its whole chromosome split by the phage enzymes. The first step in the origin of the common type of cryptic would then be a split between R and A taking place before excision of the prophage from the chromosome. Exactly how the block of genes from N to R might then be lost or how the chromosome might heal in such a manner that the $gal-bio$ distance is increased must be left vague. The difference between standard lambda and lambda *crg* might be in the "healing" process rather than in excision or end joining. We are considering always a small fraction of the survivors of high doses of irradiation. It might, for instance, happen that splitting without excision is lethal for lysogens of standard lambda but not for those of lambda *crg;* in that case, the critical difference between standard lambda and lambda *crg* could be a structural one, even if the primary event producing a cryptic phage is mediated by a phage-specific enzyme.

Whatever the basis of cryptogenicity, it seems likely that its further study will shed light on the normal processes of attachment and detachment.

GENETIC CONTROL OF ABNORMAL DETACHMENT

Other cases are known where genes effect abnormal detachment of a prophage. We have mentioned in Chap. 7 that detachment of integrase-deficient mutants can lead to defective phage formation.

The isolation of bacterial deletions that penetrate the $\phi 80$ prophage was described in Chap. 6. The first two mutants depicted in Fig. 5-2 are notable in that they contain all known phage genes, but that the yield of active phage following induction is only 10^{-4} to 10^{-5} times the normal amount. Franklin (1967) has shown that the active phage produced by such strains are highly heterogeneous. Independent isolates each carry a characteristic, hereditarily stable deletion of genes between *J* and *N*. Gratia (1967) has found that most of the active phage coming from some similar deletion mutants are transducing phages carrying the tryptophan region. Both observations can be interpreted as consequences of the absence of the *att* region. Our diagram of Fig. 5-2 embodies this interpretation. The alternative would be that unknown phage or bacterial genes involved in normal excision have been deleted.

The existence of such strains showing abnormal excision raises the question whether most transducing phages might arise, not directly from normal phage, but from a few variants. Obviously, any population of lambda lysogens will contain some fresh mutations of lambda *int* and some bacterial deletions penetrating the prophage. Their products will contribute to the phage yield formed by any lysogenic culture. Clearly, some of the transduction titer of a normal lambda lysate must be attributable to such mutants. No mutant so far reported yields enough transducing particles per cell to account for more than a tiny fraction of the total amount formed, however.

9

IMMUNITY AND ITS GENETIC CONTROL

REPRESSION AND SUPERINFECTION IMMUNITY

In Chap. 1, we defined the two main problems raised by the existence of lysogeny (*genetic* and *physiological*, respectively). Our discussion up to this point has been concerned primarily with various ramifications of the genetic problem. The physiological problem, to recapitulate, is that the virus genome is potentially capable of causing syntheses lethal to the cell, so that the cell can harbor such a dangerous guest indefinitely only if these potential syntheses are effectively repressed.

This is the physiological problem at the cellular level. Another (in principle, independent) problem arises at the populational level. In any lysogenic culture, some cells will lyse and produce phage spontaneously. If these phage could multiply on other lysogenic cells they infect, the entire culture would soon be decimated by the virus produced by some of its members. The lysogenic cell must not only be able to keep its own prophage under control but also must be immune to external infection.

Such superinfection immunity is indeed observable experimentally. In some phage-host systems, lysogeny actually results in an alteration of the bacterial surface so that homologous phage can no longer attach (Uetake and Uchida, 1959). Such lysogenic conversions are of interest in their own right, but tend to obscure the fact of greatest relevance in the present context. Many temperate phages, such as lambda, can attach to lysogenic cells and inject their DNA, but nevertheless fail to damage the host to any detectable extent.

As superinfecting phage are not blocked in penetration, they must be neutralized or repressed at some later stage. Whatever it is about the lysogenic cell which imparts this ability to counteract an entering phage genome should be sufficient to keep the resident prophage quiescent as well. In other words, the problem at the cellular level has the same solution as that at the populational level, whatever that solution may be.

Such reasoning requires that both superinfection immunity and the repression of prophage functions should be mediated by physiological factors such as diffusible repressors rather than by steric effects of prophage attachment. The evidence in the two cases is as follows:

Immunity. One explanation for immunity entertained in the early 1950s was that a temperate phage must interact with its chromosomal attachment site (its "ancestral home") in order to multiply. In a lysogenic cell, this site is blocked by the prophage, and multiplication cannot occur. This was first shown to be false with phage P2, which can occupy more than one site on the bacterial chromosome. Cells with P2 at any site are immune; conversely, the chromosomal site can be blocked by a heteroimmune related phage without causing immunity. Equivalent evidence for phage lambda came from the study of partial diploid strains, where the segregation of phage-sensitive cells from lysogenic diploids is readily observable. Furthermore, bacterial mutants from which the prophage attachment site has been deleted altogether still support multiplication of infecting phage particles (Adhya, 1968).

Repression. The prophage of a lysogenic cell can be induced to multiply and function by a variety of methods, some of which are connected by genetic arguments to the same factors determining superinfection immunity. The clearest demonstration that the prophage itself is repressed by diffusible cytoplasmic factors, however, is the induction of phage development that accompanies introduction of a lysogenic bacterial chromosome into a "nonlysogenic" cytoplasm—either by bacterial mating or by transduction. Introduction of the prophage into a nonimmune cytoplasm disrupts the stability of the lysogenic condition. No induction is observed in control experiments where the recipient is lysogenic.

GENERAL REMARKS CONCERNING GENE REGULATION

Any discussion of the immunity of lysogenic bacteria and its mutational control must be set in the context of present knowledge of gene regulation. This is an area of very active current research. We shall not attempt an extensive survey, but shall confine ourselves to mentioning the most salient general features that have emerged. The major synthetic formulation of the basic aspects of gene control was made by Jacob and Monod (1961). Further work has served for the most part to clarify some details of the fundamental mechanisms they proposed.

To understand the control of gene function, it helps first to understand the process under control—i.e., how genes act. Jacob and Monod derived some important surmises about gene function from experiments on gene control. These surmises have since been verified by more direct biochemical evidence and can now be taken as established facts.

We can start with DNA. The DNA molecule of a bacterium is divisible into discrete genes, each of which carries the genetic code for the amino acid sequence of a polypeptide chain. The genes do not function directly in protein synthesis. They are first transcribed into RNA. The "messenger RNA" molecules have a base sequence complementary to one strand of the DNA molecule. It is the messenger RNA which then attaches to the ribosomes and specifies the sequence of the polypeptide for which it codes. Messenger RNA molecules are frequently, but not always, highly unstable *in vivo*, compared to other forms of RNA, such as ribosomal or transfer RNA.

One conclusion derived indirectly by Jacob and Monod and later amply verified is that a single messenger molecule can carry information from several genes. The genome can thus be divided up into discrete regions called operons. The genes of one operon are all transcribed onto the same messenger molecule. Different operons are transcribed on different molecules.[1]

Transcription proceeds linearly, from one end of the operon to the other. It follows that any factor affecting the initiation of transcription of an operon will influence the rate of transcription of the entire operon. This will be reflected in a coordinate effect on the rate of synthesis of all those proteins coded by genes in the operon. Whenever the messenger is very unstable, this rate at any time will be determined by the rate of messenger synthesis at the same moment.

The major experimental contribution of Jacob and Monod was the discovery of specific genetic factors controlling the rate of transcription of a given operon. For the *lac* operon of *E. coli*, two such

[1] This statement goes beyond the actual data. The existence of polygenic messenger molecules is well established. The assertion that two such molecules never overlap is partly an assumption and still rests on indirect arguments. The primary definition of the operon given by Jacob and Monod was a block of closely linked genes whose transcription was under common control. As we shall see, this definition accords closely with the one given here.

factors were discovered: (1) A *regulator* gene (*lacI*), which synthesizes a diffusible protein *repressor* that inhibits transcription of the operon. (2) An *operator* region, close to the origin of transcription of the operon, which is specifically recognized by the repressor of that operon. The operon is rapidly transcribed in the presence of repressor when certain small molecule inducers are added to the medium. It was proposed that these inducers specifically combine with, and inhibit, the repressor. These conclusions were deduced from the properties of various mutants of the regulator and operator loci in which the normal behavior was deranged. The regulator gene product has since been isolated. It is a protein with a specific affinity for the inducer molecule (Gilbert and Müller-Hill, 1966).

The theory of gene control developed by Jacob and Monod had four basic premises:[2] (1) The rate of messenger synthesis is determined by the rate at which transcription is initiated. (2) The rate of transcription initiation is regulated by the action of a specific protein on a specific DNA sequence near the site of initiation. (3) The effect of this protein on initiation is negative. It is a repressor, rather than an activator. (4) Small molecule effectors can combine with the repressor. For some operons, repression is caused by the free "apo-repressor." For others, it is the repressor-effector complex that is the active repressor.

Jacob and Monod showed that, by combining the basic elements of operator, regulator, repressor, and effector in various manners, many different kinds of regulatory circuitry could be constructed, perhaps enough to account for all cases of gene regulation. Their work left open the question of whether the four premises listed above really were universally applicable, or whether the proper generalization of their discoveries might take some different form.

No good evidence has yet been adduced to challenge the validity of the first, second, or fourth premise. The possibility that in some instances control might be at the translational rather than the transcriptional level is still quite open, but there are no demonstrated examples. For the purpose of exposition, we shall throughout our discussion here assume premise (1) and equate the activity of a gene with its rate of transcription, even where the latter has not been directly determined.

Work with other systems indicates that the third premise (negative control) is not universal. There are several examples, including

[2] Not all these premises were explicitly stated by Jacob and Monod. They represent my reformulation of their theory to include some evolution in concepts that has taken place since their original paper appeared.

some from phage lambda, in which some gene regulates an operon in a positive, rather than a negative, manner. Some of these may be trivial cases where a gene codes for an enzyme that synthesizes a small molecule effector. Others are not explainable on that basis.

For example, Sheppard and Englesberg (1966) reported that the *araC* gene is necessary for synthesis of enzymes of the arabinose pathway. The significance of their findings with respect to the operon concept is best understood by a comparison of the *araC* gene and its mutants with the *lacI* gene studied by Jacob and Monod.

TABLE 9-1. *Effect of Regulator Gene and Inducer on Enzyme Synthesis*[a]

Lac I System[b]		Inducer[a]		AraC System		Inducer	
Alleles Present		Present	Absent	Alleles Present		Present	Absent
Haploid configurations	I^+	+	−	C^+		+	−
	I^-	+	+	C^c		+	+
				C^-		−	−
Diploid configurations	I^+/I^-	+	−	C^+/C^-		+	−
				C^+/C^c		+	−
				C^c/C^-		+	+

[a] For simplicity, the *lacI*s allele has been omitted from the table (see text).
[b] *Symbols*   + = high rate of enzyme synthesis
             − = little or no enzyme synthesis
            I^+ = wild type (inducible) allele of the *lacI* gene
            I^- = constitutive allele of the *lacI* gene
            C^+ = wild type (inducible) allele of the *araC* gene
            C^c = constitutive allele of the *araC* gene
            C^- = negative allele of the *araC* gene

Table 9-1 shows the pertinent facts. The wild-type alleles of both genes allow little transcription of the operon in the absence of inducer. Both genes can mutate to a constitutive allele (C^c or I^-), in which transcription occurs even in the absence of inducer. The *araC* gene can also mutate to a third type (C^-), not found with *lacI*,[3] where transcription occurs neither in the presence nor absence of inducer. The fact that deletions of the *araC* gene have a C^- phenotype, as well as the completely recessive nature of C^- mutants, indicates that C^-

---

[3] The *lacI* gene can mutate to an allele (I^s = superrepressed) that forms no enzyme under any condition. However, this allele differs from *araC*$^-$ in that it is dominant over I^+ and I^- in diploids.

is a null allele, corresponding to the complete absence of C gene function.

More revealing are studies of partially diploid bacteria heterozygous for these alleles. For the alleles of *lacI* shown here, only one combination is possible (I^+/I^-). In the absence of inducer, the diploid shows the same low level of transcription characterizing the I^+ allele. Inducibility is dominant to constitutivity. This is the classical observation from which the idea of negative control originated. The diploid behaves as though the I^+ allele makes a product that represses, and as though the I^- product is either absent or unable to repress (i.e., as though I^- were a null allele).

Precisely the same situation occurs with *araC* in the combination C^+/C^c. By the same logic, C^c might be considered a null allele. However, that is not the case. The true null allele is C^-. Both C^+ and C^c are dominant over C^-, and the C^- phenotype is no transcription. The C^+ gene product thus regulates transcription in a positive manner, even though it prevents constitutive synthesis by C^c.

Faced only with the data of Table 9-1, a student of comparative gene control would probably start with the C gene as a more general case and assume that the mutants of *lacI* corresponding to C^- have not yet been found for technical or trivial reasons. I have been unable to generate the slightest enthusiasm for this viewpoint among my friends who study the *lac* operon, however. No recessive *lac*$^-$ mutants have been mapped within the *lacI* gene. The existence of I^- mutants of the nonsense (amber) type as well as deletions entering the *lacI* gene give every indication that in this case the null phenotype is the I^- one, as Jacob and Monod suggested.

As regulation occurs at the level of transcription, a complete formulation of the mechanism of regulation must include the biochemical mechanism of transcription. An enzyme (RNA polymerase) is known that has the properties expected of a "transcriptase." It produces, on a DNA template, RNA molecules complementary in sequence to that of the DNA.

There are at least three important intracellular processes where enzymes act on DNA substrates: replication, transcription, and recombination. For all these processes, specific genes can control the sites on the DNA molecule that serve as origin of replication or transcription, or as the locus of crossing over. Enzymes catalyzing replication and transcription *in vitro* are known. Their pertinence to the *in vivo* process remains to be established. The enzymology of recombination is unknown, but genetic studies of phage lambda indicate that there is both a generalized recombinase (*red*-gene product) and a specific recombinase (*int*-gene product). From the site specificity of the *int* gene product, we have argued in Chap. 6 that recognition is an attribute of an element playing an active role in recombination.

It is not known whether all transcription *in vivo* is mediated by a generalized polymerase or whether there are specific polymerases as well. Universal negative control is most easily understood if the cell has only one kind of polymerase. The polymerase then initiates synthesis at any potential initiation site not blocked by a repressor. At the other extreme, if every operon were to have its own specific polymerase, positive control should be the rule.

The properties of the *araC* locus are those expected of the structural gene for a specific polymerase. Mutants in which the gene is absent or inactive cannot transcribe the operon. The wild-type polymerase is activated by a small molecule effector. The C^c mutants would form a polymerase active even in the absence of effector. Where both C^+ and C^c products are present, it is easy to imagine that the C^+ product in the absence of inducer can combine with its DNA substrate and block transcription by C^c.

Formally, of course, the *araC* product need not be a polymerase, any more than the *int* gene product need be a recombinase. All we know is that *araC* is an active element playing a positive role in transcription.

Present information allows no definite conclusions as to the relative importance of generalized vs. specific polymerases. Specific polymerases have been proposed for the transcription of late phage genes, mainly on the basis of positive control. We might hope that regulation would usually be the province of specific rather than generalized polymerases, but nature does not always operate according to our notions of simplicity.

The finding of positive control, in any event, should not be considered as a violation of the Jacob-Monod theory, but rather as an extension of it. The most basic feature of their theory is that a specific protein recognizes the operator and affects transcription initiation. Whether this effect is positive or negative seems of secondary importance, at least to this author.

As to the nature of the recognition region on the DNA (the operator), the situation with respect to transcription is very similar to that already described for *int*-promoted recombination. In both cases, the actual site of action can be distinguished from the recognition region, and seems to occupy a position close or adjacent to it. For the *lac* operon, this site (the *promoter*) has been located on the distal side of the operator region, indicating that the operator is transcribed, although perhaps not translated (cf. Fig. 9-1). Mutations of the promoter depress the rate of transcription, whereas mutations of the operator alter or destroy the ability to be repressed (Ippen et al., 1968).

Operator mutations differ from regulator mutations in that they show a *cis* effect. The operator exerts its influence by virtue of its con-

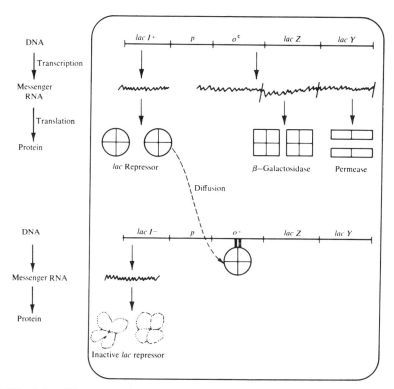

FIG. 9-1. Diagrammatic representation of control of the *lac* operon, as illustrated by a particular example. The cell shown is diploid for the *lac* region, with genotype I^+o^c/I^-o^+. Inducer is absent. The *lacI* genes are transcribed and translated, but the I^- allele gives an inactive product. The active repressor made by the I^+ gene diffuses through the cell to combine with the o^+ gene *trans* to it, thereby preventing transcription of the adjacent *lac* operon. The o^c mutant has lost the affinity for repressor characteristic of o^+, and therefore transcription can proceed from the promoter site adjacent to o^c even though repressor is present.

tiguity to the operon affected. The diploid configurations $o^c\ lacZ^+/o^+\ lacZ^-$ and $o^c\ lacZ^-/o^+\ lacZ^+$ are thus not equivalent. An illustration of this property and of the general model for control of the *lac* operon is given in Fig. 9-1.

In our discussion here, we have elected to take presence on one messenger molecule as the ultimate criterion by which two genes are assigned to the same operon. In practice, molecular measurements of messenger are frequently not directly available. The working definition of molecular geneticists employs two criteria: Two genes are in the same operon if they are controlled by the same operator, and if certain mutants of one gene show a polar effect on translation of the second.

The polar effect is due to the fact that the "nonsense" chain-terminating codons UAG and UAA apparently cause the messenger molecule to dissociate from the ribosome, thus blocking translation beyond that point. The ribosome moves along the messenger molecule as translation proceeds. If a nonsense codon in one gene is created by mutation, not only is the protein formed by that gene unfinished, but those genes distal to it on the messenger are translated at a much lower rate. This creates a pattern of directional polarity away from the promoter end of the messenger. Nonsense mutations of *lacZ* for example, produce reduced levels of the permease coded by *lacY*. Mutations in the *lacY* gene never affect synthesis of the β-galactosidase coded by *lacZ*.

In constructing their general theory of genetic regulation, Jacob and Monod drew on two examples from their own work: the *lac* operon and bacteriophage lambda. Despite the important historical role that the immunity of lambda lysogens played in the development of the operon concept, it is only very recently that much direct information on the lambda has become available. Whereas some of the work is still too recent to have received critical evaluation, we shall see that the evidence fits extremely well into the general formulation of Jacob and Monod.

GENETIC CONTROL OF IMMUNITY

We shall start by describing the experimental facts with lambda in primitive operational terms that antedate the Jacob-Monod theory, and return to their evaluation in terms of the theory shortly.

Lysogenic immunity is defined operationally by the fact that a lysogenic bacterium does not support growth of superinfecting particles of the prophage type. The immunity is specific. Among naturally occurring phages, even some that are closely enough related to recombine genetically, we find patterns such as that illustrated in Table 9-2 for phages lambda and 434. Each of the two phages can grow on lysogens of the other, but not on its own lysogens. (The terms *homoimmune* and *heteroimmune* are useful in describing this situation.)

TABLE 9-2. *Pattern of Response to Cross-infection Shown by Two Heteroimmune Phages*

Superinfecting Phage	Lysogen[a]	
	$K(\lambda)$	$K(434)$
λ	−	+
434	+	−

[a] + = phage growth (no immunity); − = no growth (immunity).

From the pattern of Table 9-2, we can define, formally, four properties of a temperate phage: (1) Ability to generate a signal. Lambda prophage must do something that renders the cell immune to lambda. (2) Ability to receive a signal. The superinfecting lambda recognizes that the cell is immune. (3) Specificity of signal generation. Lambda generates a different signal from that of 434. (4) Specificity of reception. Lambda responds to the lambda signal, not to the 434 one.

The Jacob-Monod theory predicts that properties (1) and (3) are determined by a regulator gene, whereas properties (2) and (4) are attributes of an operator-promoter region. In fact, all four properties are determined by a small region of the lambda genome that lies between genes *N* and *cII*. All mutations affecting ability to generate or respond to immunity map in this region. Crosses between related heteroimmune phages such as lambda and 434 show that both the specificity of generation and the specificity of response are determined there.

Regulator genes frequently map close to the operons they control. No very convincing explanation has been advanced for why this is so, but it is true both of *araC* and of *lacI*. So it is quite reasonable to imagine that both regulator and operator of lambda lie within one small, connected region of the genome.

If lambda is crossed with its various heteroimmune relatives such as 434, 21, etc., recombinants can be obtained that carry genetic markers from the lambda parent but have the immunity of 434, or vice versa. However, some markers, including those of the *cI* region, cannot be "crossed into" the heteroimmune stock.

Inability to recombine with the immunity determinant is not restricted to the *cI* region but applies equally to some markers in adjacent regions of the map. The exact extent of this "nonrecombinable" region for different phage pairs; e.g., some mutations can be recombined with the 434 immunity, but not with that of phage 21

(Fig. 9-2). Molecular studies of the type diagrammed in Fig. 2-5 indicate that the immunity regions of lambda and 434 are not sufficiently similar in base sequence to hybridize with each other.

In any event, this situation renders impossible any direct test for separation of the specificities of signal generation and signal reception. Crosses betweeen lambda and 434 fail to produce any recombinants that generate one immunity and respond to another (Thomas, 1964).

The immunity regions of different phages are so dissimilar that no recombination takes place between them. Is this fact of any evolutionary significance relevant to the immunity property itself? The main point to make on this subject is that recombination betweeen naturally occurring heteroimmune phages, while it does occur, is rare in all regions of the genome, and has not been specifically tested throughout the genetic map. The fact that markers distant from the immunity region can recombine with it means there are some stretches of homology located in the intervening map regions, even if there is no homology in the neighborhood of the markers themselves.

DISSECTION OF THE IMMUNITY REGION INTO COMPONENT ELEMENTS

Crosses between heteroimmune phages focus our attention on the region between N and Y. It is there that the entire specificity, both of generation and response, must lie. Because recombination in this region does not take place betweeen heteroimmune phages, further analysis depends on the study of mutations mapping within this region.

REGULATOR MUTANTS

The lambda prophage, like the *lac* operon, is controlled negatively. In mutants that have lost the ability to generate immunity, phage growth is unrepressed. They form clear plaques, free from the normal overgrowth of lysogenic bacteria.

Detailed studies have been made of clear plaque mutants in bacteriophage lambda and in the *Salmonella* phage P22. The two phages behave similarly. We shall discuss mainly the lambda results. Following infection with a clear mutant, the cells all lyse and liberate phage. There is no (or, with some mutants, very little) lysogenization. Following mixed infection with wild-type lambda and a clear mutant thereof, many cells become lysogenic. Qualitatively, ability to lysogenize is dominant over inability to do so.

REGULATOR MUTANTS

Many independent clear mutants have been isolated and mapped. Mixed infection by a pair of mutants may lead to lysogenization or not, depending on which pair is used. Some pairs can complement each other so that lysogenization is observed with the mixture but not with either mutant singly.

Such complementation tests allow subdivision of the mutants into three groups, distinguished by the fact that complementation is seen only between members of different groups. Genetic mapping shows that

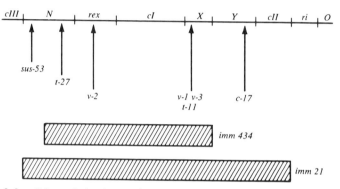

FIG. 9-2. Map of the immunity region of lambda and related phages. The genes shown have been described in Chap. 2 (Fig. 2-1), except for Y and ri. Y is defined as a region rather than as a gene. Its left boundary is the terminus of the $imm434$ region, and its right boundary is the cII cistron. ri is the locus of mutations allowing escape from replication inhibition (see Chap. 10).

Of the mutants shown, sus-53 is an amber mutant and t-27 is a defective mutant, both in gene N. The virulence markers, v-1, v-2, and v-3, the clear mutant c-17, and the defective mutant t-11 are described in the text.

The bars indicate the region of nonhomology between the immunity regions of phages lambda and 434, and lambda and 21. Markers outside of the bars can recombine with the immunity determinants. For example, a cross between $\lambda susN53\ imm^\lambda$ and $\lambda N^+imm434$ will yield some N^+imm^λ recombinants, but the cross $\lambda susN53imm^\lambda \times \lambda N^+imm21$ will not.

each group occupies a connected region of the genetic map. The three groups (cI, cII, and $cIII$) lie close to each other (Fig. 9-2).

A mutant that fails to generate immunity should form clear plaques, because the turbidity of the wild-type plaque is caused by growth of immune cells. The converse is not true, however. Mutants of the cII and $cIII$ groups can generate immunity. The frequency of lysogenization is low but not zero. The lysogens, once formed, are stable and immune to superinfection. As seen in Fig. 9-2, cII and $cIII$

mutants map outside the immunity region and therefore cannot be concerned with the specificity of immunity.

The *cI* mutants, on the other hand, show no evidence of immunity. They never lysogenize by themselves. In mixed infection with *cII* or *cIII*, they can help the other mutant to lysogenize and can themselves be carried in doubly lysogenic bacteria along with *cII* or *cIII*. The double lysogens can produce singly lysogenic segregants carrying *cII* or *cIII*, but never *cI*. The *cI* mutants are still susceptible to immunity, however. They do not form plaques on lambda lysogens.

The cII^+ and $cIII^+$ functions are therefore required for lysogenization, but only cI^+ is necessary for maintenance of the lysogenic state. This is shown more directly with temperature-sensitive *cI* mutants. These form clear plaques at high temperature and turbid plaques at low temperature. Lysogens can be made and maintained at low temperature, but heating quickly induces lysis. The *cI* gene of lambda codes for a protein product, detectable *in vitro* (Ptashne, 1967a).

This protein recognizes and distinguishes lambda DNA from that of a lambda 434 hybrid phage carrying the immunity region of 434. Echols and co-workers (1968) have recently demonstrated that an immunity-specific substance inhibits DNA transcription *in vitro*.

The temperature mutants divide the *cI* gene into two subregions. Mutants in region A are induced by a brief exposure to elevated temperature, even in the presence of chloramphenicol. Region B mutants, on the other hand, are induced only when protein synthesis at high temperature is allowed.

In the *cI* region there also is located a mutant called *ind* (non-inducible). Lysates of lambda *ind* are not induced by ultraviolet light or mitomycin C whereas those of wild-type lambda are. The *ind* character is dominant in double lysogens and also in superinfection. If a lambda lysogen is induced and then superinfected by lambda *ind*, induction is reversed. Presumably the *cI* product of *ind* differs either in quality or in quantity from the normal one.

GENETIC PROGRAM AND REGULATORY CIRCUITRY OF THE LAMBDA VIRUS

Before discussing operator mutants, we may inquire about the operons they control.

In the lysogenic cell, the amount of phage-specific messenger RNA is very small, almost negligible. What transcription there is seems to be largely of the *cI* gene itself. When the cell is induced, genes from many parts of the phage chromosome are transcribed and translated. Clearly the immunity controls, directly or indirectly, expression of the entire phage genome.

As with other large viruses, the lytic cycle of lambda phage is a complex affair. After infection of a sensitive cell or induction of a lysogen, numerous events occur according to a regular time schedule, which culminates in cellular lysis and liberation of mature virus particles. One of the goals of molecular biology has been to understand all the mechanistic details in the determination of such a sequence of events. Lambda constitutes one of the best materials for such studies, because of the wealth of genetic, physical, and (to a lesser degree) biochemical tools now available.

At the time of writing, a concerted effort is underway to understand the genetic program of lambda. No very useful purpose will be achieved by recording here the latest scraps of available information when more definitive conclusions are clearly in the making. It will suffice to present the subject in broad outline.

The most elementary, well-documented fact is that the program is determined by control at the level of gene transcription. Obviously, observable events depend not only on the time of transcription but also on the time required for gene products to interact. The important thing is that such interactions do not tell the whole story. Different regions of the genome are transcribed over different time periods, and consequently the genes in these regions are expressed at different times. As shown in Fig. 2-3, early transcription takes place in regions of the lambda molecule adjacent to the immunity region. Transcription of the *cI* gene itself and genes to the left of it proceeds in a leftward direction. Transcription of genes to the right proceeds rightward. The left end of the lambda molecule and probably also gene *R* are transcribed later, in a right-hand direction.

We now return to the question of immunity. Directly or indirectly, immunity represses transcription of the whole lambda molecule, except for the *cI* region and the adjacent *rex* gene. Experimentally, direct control can be distinguished from indirect control by superinfecting a heteroimmune lysogen with a mutant defective in the gene under test, and asking whether the superinfection induces the corresponding prophage gene to function.

The basic experimental design as devised by Thomas (1966) is shown in Fig. 9-3. Two controls were necessary: (1) Failure of a prophage gene to function might be attributed to its physical connection to the bacterial chromosome rather than to immunity. This possibility is excluded by showing that immunity will also block gene expression by a superinfecting phage homoimmune to the prophage. (2) Apparent absence of gene expression by the prophage might result from the method of assay, which measures ability of a prophage gene

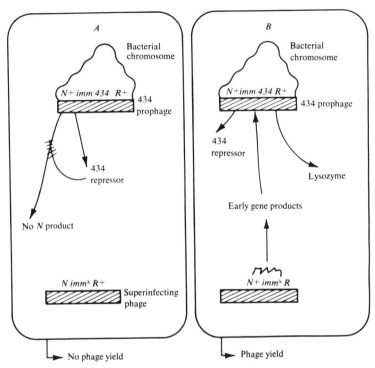

FIG 9-3. Diagrammatic representation of Thomas' superinfection of heteroimmune lysogens. In A, a 434 lysogen is superinfected with a mutant in a gene (N) that is shut off directly by the immunity. The N product is therefore not formed by either of the two phage genomes present, and the cycle cannot go to completion.

In B, superinfection is by a mutant in a gene (R) shut off *indirectly* by immunity. The other genes of the superinfecting phage (including N) will function in the presence of immunity. Their products activate gene R; gene R makes lysozyme, and a normal growth cycle follows. The absence of lysozyme synthesis in the lysogenic cell is explained on the assumption that the R gene requires activation by the products of certain genes that are repressed by the immunity.

to supply a function for the superinfecting phage. The inherent ability to complement can be estimated by inducing the lysogen prior to superinfection.

Of all the lambda cistrons tested, cistron N is clearly blocked by the immunity. *N* mutants will not grow on bacteria carrying a heteroimmune prophage with an intact N^+ gene. Cistron R mutants, on the other hand, will grow on such a lysogen, and enzymic assays of its product, lysozyme, show rigorously that the prophage gene is indeed turned on by the heteroimmune superinfection.

The effect of immunity is thus to introduce one or more very early blocks to phage reproduction and function. When these early blocks are removed by allowing the genes in question (of which *N* in one) to act, the other genes of the phage are thereby induced.

Genes *cII* and *cIII*, which control early steps in lysogenization, are directly shut off by the immunity (Bode and Kaiser, 1965). We have mentioned in Chap. 6 observations suggesting that the *int* gene that effects chromosomal attachment is likewise directly controlled.

Those genes under direct control of the immunity are therefore certainly *N, cII,* and *cIII,* probably *O,* and perhaps *P, Q,* and *int.* Other genes between *int* and *Q* have not been tested. *R* is certainly controlled indirectly, and probably *A-J* are also.

Comparing these results with the genetic map of lambda, we see that those genes under direct control are contiguous to the *cI* region. They are all genes whose function is expressed early in the lytic cycle.

The ability of certain genes such as *R* to be turned on by infection with a heteroimmune phage says that their activity is induced by prior function of other lambda genes. The early gene *N* and also gene *Q* are both necessary for efficient functioning of *R* and other late genes. Control of lambda gene function is thus effected by specific factors having positive as well as negative effects.

OPERATOR MUTANTS

So far as operators are concerned, the location of those genes under direct immunity control, plus the fact that the specificity of response to immunity is localized between *N* and *cII* lead us to postulate the existence of two operator regions, one between *cI* and *N* and a second between *cI* and *cII*. The known directions of transcription fit the idea that *N* and *cIII* could be controlled by the first operator and *O* and *cII* by the second.

The pertinent regions of the genetic map are shown in Fig. 9-2. Between *cI* and *N* on the left lies the *rex* gene, which determines ability to exclude *rII* mutants of phage T4. Between *cII* and *cIII* are the *X* and *Y* regions, whose border is defined arbitrarily as that point beyond which mutants can recombine with the 434 immunity determinant.

The most interesting mutations with respect to operator sites are *v-2*, *v-1*, *v-3*, and *c-17*.

All of these mutants have originated from virulent stocks of lambda. A virulent stock is one that can multiply on an immune host. Virulent mutants of lambda occur very rarely and turn out always to have undergone multiple mutations from wild type. A virulent stock studied by Jacob and Wollman in 1954 carried the two genetic alterations *v-2* and *v-1 v-3*, separable from each other by crossing. *v-1 v-3* is assumed to be a double mutant because the *v-3* component has been found alone. *v-3* stocks give rise to *v-1 v-3* mutants, which wild type does not (at a detectable rate). *v-1* and *v-3* have not been separated by recombination.

Physical studies show that the *cI* product binds less strongly to *v-1* or *v-2* DNA than to DNA of wild-type lambda. Complementation studies indicate that *v-3* shows constitutive expression of *O* but not *N*, whereas *v-2* is constitutive for *N* but not *O* (Ptashne and Hopkins, 1968). Neither *v-1 v-3* nor *v-2* is virulent by itself. This is expected because both *N* and *O* are directly repressed by the immunity, and both genes are required for phage growth.

The *c-17* mutant forms clear plaques. It is not virulent by itself, but the double mutant *cI c-17* is virulent.

Lambda *c-17* differs from standard lambda in two properties: constitutive transcription of genes *O* and *P*, and insensitivity to replication inhibition (Packman and Sly, 1968). The latter phenomenon will be discussed in Chap. 19. Essentially, it consists of a direct sensitivity of lambda replication, as well as lambda transcription, to the immunity. The reason lambda *c-17 cI$^+$* is not virulent is that immunity represses the *N* gene. Part of this immunity is generated by the *cI$^+$* gene of the *c-17* phage. If that source is eliminated by use of a double mutant *cI c-17*, the multiplying phage eventually titrates (or bypasses) the repressor, so that *N* can function.

The *v-1 v-3* and *v-2* mutations have the properties expected of operator mutations. *c-17* is outside the immunity region proper and therefore cannot be part of the true recognition region for the repressor.

If these conclusions are properly interpreted, they imply that the phenotype imparted by operator constitutivity for the early genes is virulence. This is not obvious a priori. It might have been the case, for example, that complete freedom from regulation on the part of these genes would so derange the whole lytic cycle that phage growth would not occur.

Some mutations with a lethal phenotype map in the *X* region, close to the *v-1 v-3* site. Mutants such as *t11* are defective, and exhibit hyperproduction of the exonuclease and β proteins. The *t11* phenotype cannot be that of an operator constitutive because (1) both the defectivity and the hyperproduction of early enzymes are recessive to wild type in mixed infection; and (2) *t11* responds to immunity. The observed enzyme hyperproduction occurs only after induction of (defective) lysogens of these mutants. That the protein hyperproduction is probably an indirect effect also seems likely from the location of the relevant genes. The proteins involved (exonuclease and β) are coded by genes lying between *b2* and *cIII*, which are separated from the *X* regions by the *cI* gene (Radding et al., 1967).

Of course, if a true operator constitutive were dominant defective and insensitive to immunity, it could not survive, either as a phage or as a prophage. Those mutants that are observed could then represent milder alterations in the operator locus, which are either defective or unresponsive to immunity, but not both.

In summary, although much remains to be understood about the properties of particular mutants, a body of information has accumulated that is compatible with the following notions: (1) The lambda phage has precisely two operator regions directly controlled by the immunity. (2) These are located close to the *cI* gene, on opposite sides. (3) The genes controlled by these operators begin to function early in the lytic cycle. (4) Late gene expression is shut off indirectly in the lysogenic cell, because of its dependence on the prior function of early genes.

TITRATION OF THE IMMUNITY

Several years ago, Bertani noticed that certain clear plaque mutants of phage P2, although unable to form plaques on lysogenic hosts, were nonetheless able to multiply on them to a limited extent in a single growth cycle. The ability to multiply on and/or kill lysogenic cells increased with increasing multiplicity of infection. At high multiplicities of the clear mutants, the lysogenic cells were no longer immune. Similar observations have been made with phage lambda.

It was hoped that this multiplicity effect might permit determination of the number of molecules of immunity repressor in the cell. The idea was that each molecule of repressor might combine with one phage particle. When the cell is confronted with many particles of a clear plaque mutant (which cannot itself generate immunity), its capacity is overtaxed. There is not enough repressor to go around, and the excess phage can therefore multiply.

Bertani (1965) has studied the multiplicity effect in some detail, in order to distinguish the two simplest alternative explanations for the multiplicity effect.

Saturation Model. This is the model we have just discussed. It says that the superinfecting phage neutralizes a limited number of immunity molecules. Whenever the number of phage in a cell exceeds the number of such molecules, lysis and phage growth result. If the number n of molecules per cell is constant and the phage adsorbs at random, then the fraction K of cells yielding phage is given by

$$K = e^{-m} \sum_{k=n}^{\infty} \frac{m^k}{k!}$$

where m is the average number of phage adsorbed per bacterium.

Breakthrough Model. Each infecting phage particle has a certain probability of independently surmounting the immunity barrier and initiating a lytic cycle. Here we expect

$$K = 1 - e^{-pm}$$

where p is the probability of breakthrough for an individual particle.

Leaving aside the mathematical details of these precise predictions, there is a simple qualitative property of the multiplicity dependence curves for the two models that is readily appreciated. The saturation model postulates a cooperative effect between phage particles. One phage by itself cannot saturate the immunity; some number n can. Therefore, even allowing for the random distribution of phage particles among cells, the effect of n phage per cell will be more than n times the effect of one phage per cell. So, even if we modify the detailed assumptions, the multiplicity dependence curve will always be concave upward at low multiplicities (curve A, Fig. 9-4). On the other hand, the breakthrough model makes the opposite prediction. There is no cooperative effect. The probability that n particles together will cause at least one breakthrough is just n times that for one particle if the probabilities are small. If they are large, it is less than n times, because sometimes more than one breakthrough will occur in the same cell. The multiplicity dependence curves will be concave downward at all multiplicities. As seen in Fig. 9-4, this is in fact the case.

The experiments are rather complex, because most of the infected cells do not lyse with the usual latent period, but rather over a period of several hours following infection, during which time growth and cell division take place. The results fit better the "breakthrough" than the "saturation" model. Whereas it may be possible to titrate the immunity by some method, it seems that this is not the proper way. Rather, immunity appears to be in excess at all multiplicities tested.

Another interesting fact is that the lysing bacteria liberate both the superinfecting and the prophage type. This is in sharp contrast to the result of Thomas and Bertani (1964) with virulent or heteroimmune superinfection (see Chap. 10).

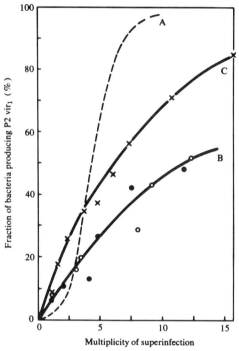

FIG. 9-4. Multiplicity dependence of phage formation in lysogenic cells superinfected with a virulent mutant of P2. Curve A is a theoretical curve calculated from the saturation model for $n=5$. Curves B and C are calculated from the breakthrough model with $p=0.06$ and 0.125, respectively. Experimental points for superinfection of C (P2) (● and ○) and C (P2 *def*) (x). [Redrawn from Bertani (1965). Used by permission of the author.]

The last finding suggests that immunity disappears from the cell by the time phage growth commences. The question raised by Bertani's experiment concerns the nature of the event that is decisive in causing this disappearance. If the disappearance depends on simple exhaustion of free repressor by stoichiometric binding to its DNA substrate, then the multiplicity dependence must be explained by complicating factors, such as heterogeneity of repressor concentration among cells. The idea that effective disappearance of repressor from the cell is determined by a single discrete event involving one phage particle provides the simplest explanation for the multiplicity curves. However, it requires the introduction of new, unknown elements to the physical picture of immunity control.

An independent estimate of the amount of repressor per cell is possible where immunity is generated by a phage genome that is neither replicating nor integrated into the chromosome. When such a

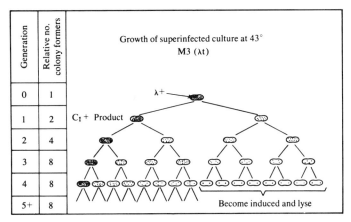

FIG. 9-5. Diagrammatic representation of persistence of heat-resistant immunity in bacteria lysogenic for a temperature-sensitive cI mutant superinfected with wild-type lambda. The fact that immunity persists two or three generations after segregation of the wild-type phage genome shows that repressor is produced in several times the minimal concentration required to neutralize an entering phage genome. [Taken from Lieb (1966). Used by permission of the author.]

cell divides, only one daughter cell retains the phage genome, but both are immune. Immunity persists for several generations after the causative gene has segregated away. This provides a minimal estimate of the number of repressor molecules per cell, as each immune cell must contain at least one molecule.

Studies of immunity in cells infected with lambda $b2$ (which is not inserted into the chromosome), or of superinfection with the wild type of lysogens whose repressor is temperature sensitive show that immunity can persist at least three generations after segregation of the phage genome (Lieb, 1966; Tomizawa and Ogawa, 1967). This indicates a minimal number of repressor molecules equal to at least four in the newly divided cell (see Fig. 9-5).

RELATION OF ATTACHMENT TO IMMUNITY AND GENE FUNCTION

Our initial classification of the problems of lysogeny into "physiological" and "genetic" implies that these are distinct problems with independent solutions. This viewpoint is fortified by the existence of mutants blocked in attachment but able to cause immunity on the one hand (like $b2$ or int) and those unable to generate immunity on the other (cI). Nevertheless, we must bear in mind that these are different aspects of the life cycle of the same organism, where interactions between initially separate processes are likely to evolve in order to permit their integration into an orderly pattern.

From what has been said in Chaps. 6 and 7, it is clear that attachment can be influenced by immunity. Immunity interferes with normal attachment, probably by repressing synthesis of a specific attachment enzyme. Following induction of a lysogenic strain, prophage is probably detached from the chromosome. When induction is reversed by superinfection with lambda ind, detachment is prevented (see Campbell, 1967b). Prophage attachment and detachment, like other genetic activities of the phage, are regulated directly or indirectly by the immunity.

On the reverse question—the effect of attachment on phage gene regulation—less is known. The fact that cells infected with lambda $b2$ reach a stable condition of complete repression indicates that attachment plays no indispensable role in establishment of immunity. Variations in gene expression between P2 lysogens whose prophages are attached at different sites have been noted.

If attachment has no influence on immunity, then some of the infected cells in which attachment takes place should subsequently lyse rather than surviving. Little information is available on this point, but the fact that lytic lysates of lambda are devoid of transducing activity suggests that chromosomal attachment and detachment during the lyitc cycle rarely if ever occur.

The relationship, if any, between attachment and immunity is central to any quantitative discussion of lysogenization frequencies. The frequency of lysogenization is defined as the fraction of infected cells that survive infection and produce lysogenic progeny. Under most circumstances, almost every survivor produces some lysogenic descendants. Frequently, not all descendants are lysogenic—indicating that prophage attachment occurred late, sometimes when the surviving cell had replicated several times after infection.

Lysogenization frequencies as thus determined do not measure prophage attachment. They measure the cellular decision (made during the first division cycle after infection) either to lyse and liberate phage or to survive as immune, potentially lysogenic cells, in which attachment will eventually occur. This decision presumably reflects the build-up of immunity in the infected cell. If attachment has no effect on immunity, it follows that lysogenization frequencies should not be altered by changing the frequency of attachment. Many proffered explanations for the determination of lysogenization frequencies have included the unstated assumption that, once prophage is attached, lysogeny should follow automatically. This assumption may be correct, but it implies an undemonstrated effect of attachment on the physiological functioning of the phage genome.

IMMUNITY OF TRANSFER FACTORS

Once the ability of colicinogeny factors and R plasmids to promote bacterial conjugation and their own transfer had been established, workers with these systems were interested for technical reasons in augmenting the frequency of transfer. It was soon discovered that cells which have recently acquired transmissible colicinogeny are far more effective as donors than their descendants, in which colicinogeny is a stable hereditary character. This renders the spread of the factor through populations highly autocatalytic.

Of the several possible explanations for this high infectivity of recent converts, the most popular at one time was that these factors resemble F in having alternatively an autonomous and an integrated state that differ in infectivity for other cells. As the evidence against an integrated state became more compelling, attention was shifted to a suggestion by Hayes that the highly infectious state represents a transient derepressed condition which ultimately disappears as the factor achieves a steady-state level of autorepression. This approach has proven more fruitful.

The arguments for physiological control of gene activity in this case are all indirect, yet convincing. As mentioned in Chap. 3, F and some R factors cause formation of specific appendages (pili) that probably serve as conjugation tubes during mating. Cells that have recently acquired the R agent show R-pili on almost every cell. In cultures where R is established, R-pili are found only rarely. Almost every cell of an ordinary F^+ culture makes F-pili, but cells carrying both R and F make very few such pili and are correspondingly less fertile.

This purely descriptive information defines the level of the problem. Conjugation is rare in established R cultures, including those that also carry F, because few pili are formed. Conjugation may fail for other reasons as well, but this is clearly a sufficient one. Pili are cellular organelles whose total complexity is not known, but which consists basically of long cylindrical tubes made of a single protein (pilin). The pilin of F- and R-pili are presumably coded by F and R, respectively. The rarity of conjugation is thus due to reduced synthesis of a gene product, not directly to the state of the element that would be transferred if conjugation did occur. The synthesis of this product is repressed in the established R culture. R can inhibit synthesis of the F product as well. Repression therefore can function in *trans*. We know rather little about the physical relationship of the relevant elements within the cell to be speaking of diffusible repressors; but a similar ignorance has not prevented application of this concept in other systems.

R agents have been classified into two categories, fi^+ and fi^- (fi = fertility inhibition), depending on whether they do or do not interfere with F function. Both types are able to promote conjugation and their own transfer. For both types, the greatest conjugal ability is evinced by recently infected cells. Both types cause synthesis of specific pili; fi^+ pili are similar, but not identical, to F pili, whereas fi^- pili seem to be different.

F is unusual among natural transfer factors in that pilin synthesis is normally derepressed, so that conjugation takes place at high frequency. Derepressed mutants of R agents have been isolated (Meynell and Datta, 1967).

The relationships between F, fi^+R and fi^-R are similar to those between phages with different immunity specificities. The fi^+ and fi^- types are analogous to two heteroimmune phages like lambda and 434; F behaves as a derepressed mutant of fi^+, analogous to a clear (*cI*) mutant of lambda.

The repressor that distinguishes fi^+ from fi^- and F agents is identified by its effect on pilus formation. Multiplication of transfer factors seems to be controlled also by repressors, but these must be distinct from those controlling pilin synthesis. The argument is again indirect: it is difficult to maintain an autonomous F agent in an Hfr culture. The presence of integrated F somehow excludes autonomous

F. (We shall assume that this exclusion results from repression; alternatives will be discussed in the next chapter.) But fi$^+$ R factors can readily be transferred and maintained in Hfr strains, where they inhibit production of F pili. The specificities of control of multiplication and of pilus formation are thus obviously distinct: F interferes with multiplication of F but not of R, R represses pilus formation both by F and by R.

CONCLUSION

In summary, we can say that there is good evidence for the idea that control of gene function in bacterial episomes, as in bacteria, is mediated by gene-specific diffusible substances. One of the ways that one species of episome differs from another is in the specificity of such substances, of which a diversified array are found among the limited number of natural isolates already examined.

Control of gene function in phage lambda is a complex process. Increasing evidence is accumulating that this complexity depends on the interactions between various elements of its regulatory circuitry. The basic genetic elements themselves are simple regulator genes and operators of the type already characterized in bacteria.

10

AUTONOMOUS REPLICATION

An episome, by definition, can multiply independently of the host chromosome when autonomous. This raises two questions: (1) How does it multiply autonomously? (2) What prevents autonomous multiplication when it is integrated?

These questions relate to the general theory of replication control, in the development of which episomes have played a major role.

The general problem of replication control began to come into focus about 1960. Most ideas about macromolecular synthesis propounded during the previous decade envisaged an initially empty template onto which subunits gradually accumulated in their proper positions, with polymerases diffusing about, randomly encountering their substrates. That picture has proven entirely wrong. All known syntheses of linear molecules on templates proceed in order, starting at one end and terminating with the other. Synthesis of the circular chromosome of *E. coli,* for instance, generally proceeds along the structure from a single initiation point (Cairns, 1963).

It is not yet certain whether this initiation point is unique, but there is increasing evidence that it is. At any rate, some effective mechanism must assure that synthesis ordinarily is not initiated at many points on the same structure at the same time.

This suggests that the decision as to whether or not a particular stretch of DNA will be replicated at any given time does not depend especially on its own physical characteristics, but rather on its status as part of a larger structure on which synthesis can be initiated. Once replication has started, all genes attached to the initiation site are replicated indiscriminately. Conversely, a loose fragment of DNA detached from the chromosome should in general not be replicated because it will lack an initiation site.

For example, the *lac* region of *E. coli* can exist either at its normal chromosomal location or as part of the transducing phage 80 *dlac*. When it is part of the chromosome, it replicates at the same rate as the chromosome; when it is part of phage 80 *dlac,* its replication rate is determined by the phage 80 moiety and may be either faster or slower than that of the bacterial chromosome, depending on circumstances. Replication of the *lac* region is then contingent not on what it is but on what it is connected to. (This statement seems firmly grounded from the biological results; to my knowledge, a direct physical measurement of *lac* gene synthesis following induction of normal phage 80 is not available.)

If it is true that a loose DNA fragment lacking an initiation site fails to replicate, it follows that any element such as a virus, a plasmid, or an episome, that can replicate autonomously, must have an initiation site of its own. An initiation site and its appended genetic material is termed a *replicon,* a unit of replication (Jacob et al., 1963).

When different replicons are in the same cell, there may be differential replication of one with respect to the other. For example, in a phage-infected cell, the phage genes may multiply while the bacterial chromosome does not. At some level or other, the phage genes respond differently than the bacterial chromosome to some aspect of the phage-infected cell. From what we have said above, this response is something that applies to the replicon as a whole: If conditions favor phage replication, for example, the lactose genes associated with phage material in a phage 80 *dlac* particle will be replicated along with the phage genes. So the determinant of response to the environment is localized on the genome. It is in the phage moiety rather than the bacterial moiety of phage 80 *dlac*. Presumably, within the phage moiety itself there are some parts that do, and others that do not, influence the response. That portion of the replicon which determines how the environment influences replication is termed the *replicator*. The replicator might in principle comprise one or more than one region of the genome, and might or might not include the initiation site.

Environmental factors specifically affecting replication of a given replicon might in principle include products coded either by the replicon itself or by another replicon, and might function in either a positive or a negative manner. A factor exerting a specific positive effect on replication will here be termed a *stimulator*. The stimulator might function either in the initiation or the continuation of replication. A stimulator coded by the replicon itself would correspond, operationally, to the *initiator* of Jacob and co-workers. The latter term should apply only to factors whose effect is on initiation.

Phage lambda can elaborate factors with both positive and negative effects on phage growth. The positive factors are the products of the "early genes" N, O, and P. The specificity range of the early genes is not known with precision. Enzymic machinery for host DNA synthesis is obviously present already in the cell prior to infection, yet lambda replication is contingent on early gene function. This suggests some specificity of these genes for lambda DNA. Complementation tests between related phages show that some early functions of phage lambda, for instance, can be supplied by genes of phage 80 (Franklin et al., 1965). Complementation between more distantly related phages cannot in general be studied because of exclusion (a phenomenon that may well depend on replication control).

The immunity repressor apparently exerts a negative control directly on phage replication, as well as indirectly by repressing early gene function.

Thomas and Bertani (1964) studied growth of virulent mutants of P2 and heteroimmune variants of lambda on immune strains. Whereas the superinfecting phage multiplies, the prophage does not. Furthermore, in mixed superinfection with an immunity-resistant and an immunity-sensitive phage (for example, in a lambda lysogen simultaneously superinfected by lambda and 434), only the immunity-resistant type multiplies. The failure of the prophage to multiply is thus not due to its physical connection with the chromosome but rather to the fact that immunity persists.

Moreover, the effect of immunity cannot be merely to inhibit synthesis of diffusible protein molecules. As the immunity-resistant phage is multiplying and maturing, all such molecules necessary for phage growth must be present in the cell. This leaves open three possibilities: (1) immunity causes a direct block to phage replication, not just an indirect block through inhibition of function. (2) Some important early phage genes have nondiffusible products that are synthesized in spatial juxtaposition to the phage genome. (3) There are some early *immunity-specific* products necessary for growth.[1] This possibility might seem remote for the P2 experiments, which employed virulent mutants differing from wild type in some early step; but until we really understand the mechanism of virulence, it remains formally possible.

Genetic control of this immunity-specific replication is under

---

[1] For example, suppose there were, within the region of nonhomology shown in Fig. 9-2, a gene whose product recognized a specific site (also in that region) and initiated replication there. The lambda initiator might fail to recognize the 434 replicator site.

current study (Dove, 1968). Mutations that abolish replication inhibition map at the *ri* locus (Fig. 9-2). The *c-17* mutation (discussed in Chap. 9) likewise does not show replication inhibition. Although the inhibition is immunity specific, these mutants can recombine with the determinants of immunity specificity. These mutations might in principle either reverse replication inhibition (by altering its site of action) or circumvent it (by creating a new site that renders the normal one unnecessary).

The section of the lambda chromosome around the *c* region thus has some structural effect on replication. However, we recall from Chap. 8 that a lambda *db* missing the gene block from *N* through *R* can still reproduce.[2] The same is true of lambda *dg*'s missing genes *A* through *J*. This means that if any region of the lambda chromosome constitutes a unique site of replication initiation, then either (1) this region must be near either the ends of the phage chromosome or *att* or else (2) the ability of wild-type lambda to promote multiplication of defective lambda can be mediated by direct physical interaction between the two genomes rather than by diffusible substances active in *trans*. The recent isolation of a lambda *dg db* containing neither *N-R* nor *A-J* (Kayajanian, 1968) suggests that even the latter possibility would require specific recognition of special lambda regions at the ends of either the phage or the prophage chromosome (since these two regions are the only sections of lambda-specific DNA that this transducing phage should have).

If one were designing an organism with paper and pencil, one might think that the simplest way to make each element initiate its own replication would be to have it elaborate a specific polymerase that recognizes some unique component of the element itself. Surprisingly, in the one case where good information on the mechanism of initiation specificity is available, the result is just that simple. The RNA virus $Q\beta$ elaborates an enzyme capable of replicating $Q\beta$ RNA. It recognizes a particular site or region near the end of the molecule (apparently, actually, a configuration caused by interaction of the two ends) and proceeds to replicate the entire sequence (Haruna and Spiegelman, 1966). We should not ignore the possibility that other systems of replication control may be equally straight-forward.[3] An analogous possibility has been mentioned in Chap. 9 concerning con-

[2] This conclusion is still tentative. The existence of this lambda *db* is clearly documented. Its ability to replicate autonomously (in the presence of helper phage) is still under study, but seems indicated.

[3] Actually, $Q\beta$ replication is not quite *that* straightforward. The "specific enzyme" is actually an enzyme complex, consisting of one host-specific and one virus-specific component.

trol of gene function. In regulation of function, as of replication, both positive and negative controls are observed. It is perhaps historical accident that the two have been given different emphasis in the two situations.

SEGREGATION OF EPISOMES AT CELL DIVISION

Whereas the autonomous sex factor F (or its associated bacterial genes in an F' agent) is genetically independent of the bacterial chromosome in crosses, a physical connection between the two structures is demonstrable *in vivo*. If labeled cells are transferred into nonlabeled medium, it is found that, through numerous divisions, labeled chromosome and labeled episome are partitioned to the same daughter cell (Cuzin and Jacob, 1967a). This suggests they are connected to each other. The connection might be indirect, by attachment of each to a common structure, perhaps the cell membrane. Several weak reasons implicate the membrane: (1) DNA synthesis, *in vivo*, is associated with the membrane fraction (Ganesan and Lederberg, 1965). (2) The membrane obviously takes part in cell division, and it is reasonable that it should be materially partitioned in a manner compatible with the genetic results. (3) Electron micrographs suggest that DNA is connected to the membrane (Jacob *et al.*, 1963). (4) Conserved units of the cell surface and conserved DNA strands are partitioned jointly at cell division (Chai and Lark, 1967).

The observed relationship between F DNA and bacterial DNA is quite similar to that between chromosomes of eukaryotes. The nonrandom mitotic partitioning of labeled chromosomes suggests that a set of DNA strands synthesized at one division cycle remain forever after bound to a common structure, perhaps the spindle (Lark *et al.*, 1966).

In both cases, caution is desirable in equating an element postulated on genetic grounds with a structure observed cytologically. In what follows, we shall try to stick to a formal genetic description of results without reference to the nature of the elements invoked. It is helpful to have an idea of what they are in the back of our heads, but this idea must not influence the logic of the analyses. In fact, very few of the conclusions that can be drawn from experimental work to date depend in any way on whether the attachment organ is really the membrane or something else.

The actual identification of a visible structure with a genetic abstraction can be a long and arduous task. For example, it has been known for some years now that the structural protein of the mitochondrion is coded by extrachromosomal genes. It has long been suspected that the mitochondrion has genetic properties. Fairly recently, mitochondrial DNA has been isolated and characterized. But at present writing there is no evidence whatever that the mitochondrial protein is coded by mitochondrial DNA. A formal genetic description of results remains the only alternative to a hopeless mixture of hard facts and soft thinking.

INTERFERENCE BETWEEN EPISOMES

One evidence for negative control of episomal replication is the specific effect of immunity shown by Thomas and Bertani for phages lambda and P2. A second is the interference between related episomes, especially those such as F where indefinite autonomous multiplication does not damage the host.

For example, when an F^+ cell becomes Hfr, it ceases to be F^+. It no longer transmits F rapidly on contact. It accepts only rarely F' agents introduced into it and selected for the bacterial genes they carry. Likewise, when two F' agents carrying different genes are introduced into the same cell, they generally segregate from each other in subsequent cell divisions. The last observation is expected in the sense that the number of F particles per nucleus in an ordinary F^+ or F' cell is low—on the order of one.

Two explanations have been proffered for these interference effects: (1) F elaborates a repressor that specifically inhibits F multiplication. (2) F attaches to a specific site or structure in the cell. The number of such sites per cell is seldom more and never less than one. Preemption of this site by one F genome excludes any other from occupying it, except by dispelling the resident F from the site.

Each of these hypotheses generates complications. If control is by repression, then stimulators and repressors must achieve such a delicate balance that F is allowed to replicate once per division cycle, but not further. It seems necessary to assume that either the F agent or the stimulators and repressors acting on it are affected by the cell division process itself; given such an effect, the formal necessity for postulating a specific repressor becomes less clear.

On the other hand, the idea of a specific F site raises difficult questions regarding the origin of the site. The information for synthesizing a specific "F site" might derive from the host chromosome, the F agent, or the site itself. Host mutants unable to harbor F can be isolated, but the nature of their genetic defect is unknown.

If the information for the specific site is derived from the F agent, we are essentially back where we started. In the absence of a specific repression system, two F agents should synthesize twice as many sites as one. If we postulate repression, then the "site hypothesis" no longer constitutes a valid alternative to the "repression hypothesis" as a mechanism for exclusion.

The same difficulty is raised even more emphatically by the notion that the sites themselves have genetic properties, that they duplicate at the time of cell division. In this case, if we postulate a specific site for each replicon, we must admit that this site itself constitutes a second "replicon," and we are left with the problem of how *its* replication is controlled.

This discussion is of course not directed against the idea that replicons in general or F in particular attach to the membrane or some other structure. A close association of F with the cell surface is suggested by its function as a transfer agent, and the site of attachment is likely to contain protein coded by F itself. The question is whether the postulation of specific sites is necessary or even helpful in explaining the interference observed between two F particles in the same cell. I find it easier to assume a specific repression than a specific, preexisting cellular site; but the main point to make is that we are still lacking a precisely formulated model for replication control.

Of those Hfr cells infected by F′, a small percentage "escape" the exclusion and engender cell lines permanently perpetuating both replicons (Cuzin, 1962.) The entering F′ may in this case be integrated into the bacterial chromosome (Maas, 1963). There is no convincing demonstration that F ever replicates autonomously in an Hfr cell.

The existence of cells carrying more than one F agent has made possible the demonstration that F resembles lambda in having genes whose products are required for its own replication. Certain mutants of F are unable to replicate at 43°C and are lost if the cell is maintained at that temperature. However, wild-type F in the same cell can supply the missing function(s), and both F agents can then be perpetuated (Jacob et al., 1963).

The indirect nature of the arguments for independent replication of transfer factors should be stressed. If there are specific sites for F replication, it cannot even be excluded that these might be chromosomal sites, and that "autonomous F" is integrated into the chromosome at the time of replication. Knowing that the lambda episome makes a specific enzyme for inserting and excising prophage, we must accept the possibility of an episome that is excised from its normal insertion site so readily that insertion is not demonstrable at all by the usual operations. On this view, Hfr cells would differ from F^+ by virtue of involving rare, unusual insertions; the "replication-deficient" mutants of F could then actually have suffered aberrations of the normal excision and insertion mechanisms; and the normal mode of F synthesis would be by means of host enzymes operating on the host chromosome in the usual manner.

CHROMOSOME REPLICATION AND BACTERIAL MATING

The replicon hypothesis implies that every independent genetic element in a cell replicates from a fixed initiation point. The bacterial chromosome itself then must have one (or possibly more) special point at which replication can be initiated, and must establish its own system of regulation operative at this special point.

Inserted episomes complicate the picture. Each episome is potentially independent and therefore has its own replication system, including an initiation point. Insertion creates a chromosome with two potential initiation sites—that of the episome, and the one(s) already in the chromosome. How do the pertinent regulatory elements so interact as to produce orderly replication and division in this case?

Not only is the mechanism of this interaction unsolved, but there is even some disagreement as to what its outcome is. Nagata (1963) reported that, in Hfr chromosomes, the site of F insertion acts as origin of chromosomal replication. Other investigators (Berg and Caro, 1967; Abe and Tomizawa, 1967), using different methods, have not corroborated Nagata's finding. There is growing evidence that *Escherichia coli* has a unique point of replication initiation (Lark, 1966), which functions in Hfr as well as F^- strains.

Differences among strains have also been observed, however, and it is conceivable that the initiation site, unique in a given strain, may be located elsewhere in another, even though the strains all stem ultimately from the same ancestral K-12 stock. Nagata's result might hold in some genetic backgrounds and not in others.

An added difficulty in incorporating Nagata's result into current thinking stems from the fact that the direction of replication he observed was opposite to that of marker transfer during bacterial mating. According to the presently favored model for conjugation (Jacob *et al.*, 1963), inserted F serves as origin of replication at the time of mating, and the process of transfer requires concomitant replication —which therefore must proceed in the same direction as chromosome transfer.

The relation between replication and transfer is still unsettled. The DNA transferred in mating is frequently composed of one new and one old strand (Ptashne, 1965b; Gross and Caro, 1966). This is to some extent expected from the amount of synthesis occurring during the mating period, whether or not the two processes of transfer and synthesis are connected. Bonhoeffer (1966) finds that cells carrying a thermosensitive mutation for DNA synthesis can function as donors but not as recipients at nonpermissive temperatures. This would suggest that at least some of the replication required for transfer takes place in the recipient; and unless the F replication system can replace the host genes required for normal replication, all of the replication would go on in the female.

Those mutants of F that are unable to multiply autonomously at 43°C also produce, when integrated, Hfr strains with reduced transfer ability. This indicates that the F replication system performs some function necessary for chromosome transfer. Whereas these mutants can multiply autonomously at 30°C, reduction of transfer ability occurs at both temperatures (Cuzin and Jacob, 1967b). This suggests that there is an alternative mechanism allowing autonomous F replication (but not chromosome transfer) at 30°C that does not require these F genes.

RELATION BETWEEN MODE OF REPLICATION AND STATE OF ATTACHMENT

In Chap. 1, I tried to document the assertion that replication can take place in two alternative states—autonomous and integrated. Ability to replicate in these two states is part of the definition of an episome.

In Chap. 1, we posed the question of two states of replication in operational terms. At this point in our discussion, we may ask whether, conceptually, this distinction between autonomous and integrated states confounds two separate issues (1) physical connectedness to the host chromosome and (2) the nature of replication control.

In other words, if it is true that each episome can form its own replication system, what is the connection between activity of this system and insertion into the host chromosome? If episome-mediated replication is an exclusive attribute of free episomes and host-mediated replication pertains only to integrated episomes, why is this so?

The second part of the latter question has a ready answer: The host system replicates the host chromosome along its entire length.

When it encounters an integrated episome, it replicates it, too. It knows of no way to bypass the episome. (If it did, the integrated episome would not be observed.) A free episome in the same cell is not replicated, because it is not part of the chromosome. This is precisely what happens when a lysogenic cell is superinfected with another phage of the prophage type. The second phage genome neither multiplies nor decays, but is simply diluted out with growth. Whether some other free elements might be replicated by the host system is another question.

Why autonomous replication does not occur in the integrated state is harder to explain. Some episomes, such as lambda prophage, pose no problem. The lysogenic cell is immune. Therefore the specific stimulators are not found, and lambda-directed replication cannot occur. If the normal situation is deranged in such a manner that immunity is lifted and excision is prevented, *in situ* replication apparently can be induced (Dove, 1967; Pereira da Silva *et al.,* 1968).

Episomes like F are harder to understand, because here autonomous replication can go on indefinitely side by side with host chromosome replication. Yet when F is integrated, replication of the complex (bacterial chromosome + episome) proceeds as though the F system had now been subordinated to the normal replication system of the host. Integrated F, furthermore, strongly inhibits establishment of a free, autonomous F introduced into the same cell.

This problem, like the interference between episomes in the autonomous state, is one to which no attractive solution has been proposed.

INTERMEDIATE STRUCTURES IN REPLICATION

At present writing, the physical nature of the replicative form is not known with certainty. Much experimental work on the subject is too recent to evaluate with any perspective. In cells infected with lambda, for instance, there is evidence for linear molecules where the usual single-stranded ends are "healed," very large rapidly sedimenting molecules (perhaps linear concatemers of the lambda genome), normal lambda molecules, and circular molecules with one or both strands covalently linked into a circle. The relation of these forms to each other, to lambda replication, and to lambda biology in general should soon become understandable.

11

JOINING AND SEPARATION OF ENDS

Insertion of lambda according to Fig. 5-1 requires two primary events—crossover between *att-PP* and *att-BB,* and joining the ends of the lambda chromosome to make a circular phage DNA molecule. Crossover between *att-PP* and *att-BB* was originally conceived as taking place by a known mechanism (genetic recombination) involving no novel elements. As described in Chap. 6, it turns out to be a sophisticated process catalyzed by a specific recombinase. The mechanism of end joining, on the other hand, was not specified in the original model. We had no idea as to how it might happen. I was excited at the time by the newly discovered circularity of the T4 genetic map, and by some ideas of Frank Stahl's that all natural DNA molecules might be circular. The only way to make lambda a circle was to join the ends, so, on paper, I just joined them. The pertinent physical data came later.

When DNA is extracted from free phage particles, the molecules can be either linear or circular, depending on conditions of extraction. The two ends of the DNA molecule can stick to each other or to the ends of the other molecules, producing either circles or polymers (Hershey *et al.,* 1963; Ris and Chandler, 1963). The two sticky ends are complementary, not identical (Hershey and Burgi, 1965). The linear molecules can be split into half molecules; two fractions, physically separable and genetically distinguishable, are found. The molecules of one fraction will stick to those of the other fraction, but not to one another.

The conditions under which the transition from linear to circular occurred were similar to those used for melting and reannealing DNA molecules. Direct evidence that the lambda ends are complementary single-stranded regions came from enzymic studies (Strack and Kaiser,

1965). The complementarity can be destroyed by DNA polymerase, which makes the ends double stranded; but it reappears when the single-stranded state is restored by exonuclease digestion. The length of the single-stranded region is about 20 nucleotides. From the speci-

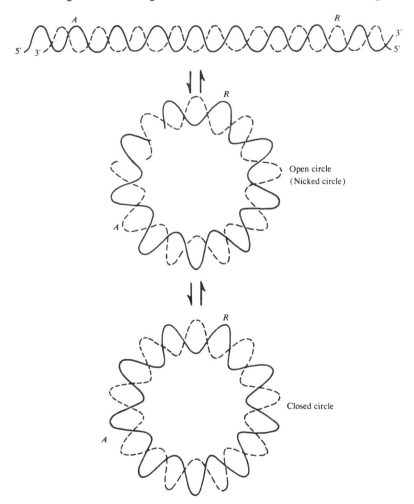

FIG. 11-1. Spontaneous annealing of single-stranded ends of bacteriophage lambda and subsequent covalent closure of the ring

ficity of the exonuclease, the complementary regions must comprise the 5' ends of their respective strands. Phages 424, 434, and 21, which recombine genetically with lambda, will form end-to-end dimers with lambda; whereas another phage (186) which forms dimers with itself will not dimerize with lambda (Baldwin *et al.*, 1966).

We can thus construct a plausible picture, well supported by physical evidence, of the transition between linear and circular molecules (Fig. 11-1). The change from linear molecule to nicked circle is a spontaneous physico-chemical process that takes place without help of enzymes.

What role, if any, does this observable test-tube phenomenon play in the biological process of end joining? It seems unlikely to be the complete story. If the nicked circle of Fig. 11-1 were incorporated as such into the bacterial chromosome, the bacterium would either have to seal it or find a way of reproducing the nicks from generation to generation. It is not too hard to imagine that a strand interruption is copied as such, but to replicate a staggered configuration of single-strand nicks requires that the daughter strand reproduce the interruption of the identical, rather than the complementary, parent strand. This is not impossible, but it is simpler to assume that "nicks" become sealed by phosphodiester bonds.

The closed circles thus generated, diagrammed at the bottom of Fig. 11-1, have been observed in lambda-infected cells. They are unable to separate into single strands like the linear molecules or open circles. No direct evidence is available as to whether these closed circles are precursors of prophage.

Closing the ring thus requires two successive steps, as shown in Fig. 11-1. The first step, from linear molecule to nicked circle, can be described in functional biological terms as a recognition process. It facilitates the discrimination by phage ends between their own complements and other DNA they encounter. It is spontaneous. The second step, whereby nicks are sealed, involves formation of covalent bonds and is probably enzymic. An enzyme from *E. coli* will catalyze this step *in vitro* (Gellert, 1967).

We cannot specify the stage of lysogenization at which sealing occurs. It might or might not function in the lytic as well as the lysogenic cycle. In any event, sealing must be reversed: Lysogenic cells can be induced to form mature particles. In these, the original linear structure is reconstituted. This suggests a specific "nicking" enzyme to make two nicks at the required places. The enzyme needs a high degree of specificity for the base sequence of the nickable region. In this respect it resembles the integrase enzyme, which recognizes the *att* region of the phage. These two regions constitute singularities on the circular phage map not for physical reasons but for biological reasons: specific enzymes can recognize them and act on them.

The time that nicking occurs cannot be specified. It might take place before prophage is excised from the chromosome, it might happen as late as maturation. All this depends on the mechanism of phage replication, which is still not completely worked out.

The idea that the phage makes a nicking enzyme as part of its normal life cycle has interesting ramifications. Lambda might then be regarded as a professional chromosome breaker, because its own life cycle entails the orderly occurrence of breaks and joins in its DNA. When the virus genome is built into the host chromosome of a lysogenic bacterium, a break in the virus genome effects a disconnection of

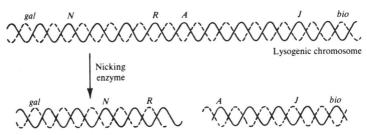

FIG. 11-2. Possible splitting of the lyosogenic chromosome by the "nicking" enzyme, as postulated by Campbell and Killen (1967).

host genes as well. In becoming lysogenic, the bacterium runs the hazard not only of destruction due to phage induction and growth, but also of chromosome breakage at the site of prophage insertion. We might expect that conditions could be found where the "nicking" reaction would take place but the integrase reaction would fail, splitting open the bacterial chromosome, as illustrated in Fig. 11-2.

Several possible examples are known. The first is the formation of the "cryptic" prophage, as described in Chap. 8. Here we must suppose that the primary splitting is followed by loss of the *N-R* region and eventual rejoining. A second is the experiments of Eisen and collaborators (1968*a,b*), where induction of certain early mutants of lambda can lead to chromosomal damage at the prophage attachment site. The latter observation is equally explainable by failure of a repair step in the recombinational event causing prophage excision.

Chromosomal splitting at the site of insertion, while hypothetical for lambda, is one of the primary characteristics of F. This is precisely what happens in bacterial mating. The closed chromosome is converted to an open one, with F-specific material apparently at both ends. There is good evidence (Chap. 10) that this opening process is connected with replication of the donor chromosome. This does not alter the fact that at least one of the two strands of the original chromosome must be cut for genetic transfer to occur. The relation of nicking enzymes to lambda replication is unknown and could be identical to that with F.

Breaking of corn chromosomes at the site of controlling element attachment is well known, and seems to be under careful temporal and genetic control.

Obviously any or all of these examples might have different explanations from the one suggested here. The important point is that the lambda phage must have a specific mechanism for recognizing a particular DNA sequence and cutting it open. It is unlikely that lambda is unique. Probably other viruses and episomes do similar things. Perhaps they all do. The system is ordinarily carefully regulated, but there must be accidents where the regulation breaks down, or where the system encounters a base sequence corresponding to its substrate in an unexpected place, as in the host chromosome. It therefore seems reasonable to approach any system in which episomes or viruses induce specific breakage with this possibility in mind.

12

POLYLYSOGENY

A bacterial strain can be lysogenic for more than one prophage. The several prophages may belong to different species, like lambda and P2, or they may be variants of the same phage, like P2 and P2c. If they are variants of the same phage, they may still be located at distinct chromosomal sites. This is usually the case for P2. On the other hand, two or more lambda prophages can be very close to each other, with no known bacterial genes intervening.

Starting from Fig. 5-1, we can guess what the gene arrangement in such double lysogens might be. If we take a lysogenic chromosome where the phage markers are arranged in a given order, say ABCD, a second phage (abcd) can lysogenize in the same way as the first, giving one of the two orders ABCDabcd or abcdABCD. Interstitial addition within the prophage should also be possible, giving AbcdaBCD, ABcdabCD, ABCdabcD.

If bacteria are mixedly infected with phage lambda and a mutant thereof, many of the stably lysogenic survivors harbor both prophages. Multiple lysogeny is not at all rare. It follows that many "single lysogenic" stocks probably carry more than one copy of the prophage. This circumstance may well have confused or complicated experiments on prophage structure.

PROPHAGE LOSS FROM DOUBLE LYSOGENS

Given a double lysogen formed by two prophages differing for several markers, one problem is to determine their gene order. In principle, this can be done by bacterial crosses. In practice, such analysis is fraught with complications caused by unequal crossing over, zygotic induction, etc. A more generally useful method has been to examine monolysogenic segregants from the double lysogen. These can result from intrachromosomal recombination, as diagrammed in Fig. 12-1. Crossing over will cause ejection of one (circular) phage genome, with retention of the reciprocal genotype. Monolysogenic segregants could also arise from interchromosomal recombination with the same formal results (cf. Fig. 4-1).

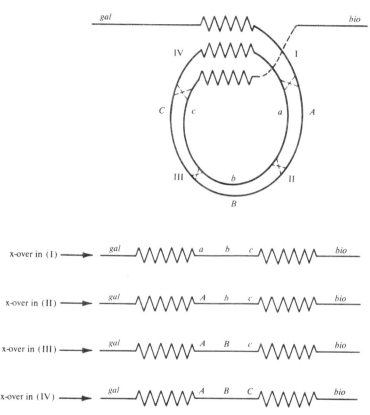

FIG. 12-1. Segregation of monolysogenic derivatives of double lysogens by intrachromosomal recombination.

If the two prophages differ in three markers, four classes of segregants can arise by single crossovers. If the gene order is that of Fig. 12-1, these classes are ABC, ABc, Abc, abc. Furthermore, these four classes can arise by single crossovers *only* from a double lysogen with the order ABCabc. If the order were, for example, AbcaBC, the four classes should be Abc, AbC, ABC, aBC. The prophage gene order can thus be unambiguously determined.

This treatment assumes that single crossovers will always be more common than multiple crossovers. This is expected if the four regions of Fig. 12-1 are of comparable length. If one region is much shorter than the others, however, this expectation may fail. A triple crossover in regions *I, II*, and *III*, for example, might be more common than a single in region *IV*.

I shall develop here a general procedure, valid for any number of markers and independent of the relative lengths of the various regions. It is also free from any specific assumptions concerning genetic distances, interference coefficients, etc. These quantities may be of interest, but to determine gene order we want first a qualitative procedure dependent on as few assumptions as possible. We assume the validity of what has elsewhere (Campbell, 1968) been called the "principle of progressive rarity," i.e., no multiple crossover type should be more common than any of its component singles or multiples of lower degree. Thus a triple in *I, II, IV* is permitted to be more common than a single in *III*, but not more than a single in *I*. The only other assumption is that the double lysogen carries exactly two complete prophage genomes inserted in tandem in the bacterial chromosome.

This assumption is sufficient to determine gene order by the following procedure (illustrated for the three-factor cross, but easily generalizable). The primary data from a segregation experiment consist of the frequencies of eight genotypic classes of monolysogenic segregants:

$$f(ABC)+f(ABc)+f(AbC)+f(Abc)+f(aBC)+f(aBc)+f(abC)+f(abc)=1$$

We examine first the frequencies of the individual markers

$$f(A\cdot\cdot)=f(ABC)+f(ABc)+f(AbC)+f(Abc)=1-f(a\cdot\cdot)$$

$f(\cdot B\cdot)$ and $f(\cdot\cdot C)$ can be defined similarly.

Now consider the six numbers $f(A\cdot\cdot)$, $f(a\cdot\cdot)$, $f(\cdot B\cdot)$, $f(\cdot b\cdot)$, $f(\cdot\cdot C)$, $f(\cdot\cdot c)$. One of these numbers will be bigger than the others. (Two of them could be exactly equal, but in that case, the analysis is even easier.) Suppose the biggest number is, for example, $f(a\cdot\cdot)$. What does this mean? It means that the amount of recombination between *a* and the end of the (double) prophage closest to it is less than that between any other marker and either end. In other words, the gene order must be $a\cdot\cdot\ A\cdot\cdot$, and the *A* locus is closer to the end of the prophage than is the *C* locus. This is directly deducible from the principle of progressive rarity.

Now we must place *B* and *C*. We first note from Fig. 12-1 that, where the order is ABCabc, any segregant carrying gene *a* must have undergone crossover in region *I*. Similarly, in our example, where the order is $a\cdot\cdot A\cdot\cdot$, any segregant carrying A must have crossed over between a and the left end of the prophage. As a single in this region will be more common than any multiple derived from it, the most common class of segregants carrying A will tell us which alleles of B and C are distal to it. Suppose, for example, that this class is Abc. The gene order must then be $a(BC)A(bc)$. To find out whether b or c is to the right, compare $f(\cdot b\cdot)$ and $f(\cdot\cdot c)$. If $f(\cdot b\cdot)<f(\cdot\cdot c)$, then the order is $aBCAbc$.

There are alternative methods for finding the gene order in the double lysogen, but those that are valid are logically equivalent to this one. Our treatment assumed only the principle of progressive rarity and tandem insertion. We were not required to know in advance the order of the prophage map in the single lysogens. This order is deducible from the segregation data. Likewise, we required no knowledge of the coupling relations of the markers in the two phages from which the double lysogen was derived. Knowing the parental couplings, we can deduce the integration event that produced the double lysogen.

For example, suppose the prophage order of the double turns out, as above, to be $aBCAbc$. If this double was made by superinfection of $K(abc)$ with ABC phage, then the integration crossover occurred between a and b.

Now suppose we superinfect $K(abc)$ with ABC phage and examine many different, independent double lysogens produced. From the segregation pattern of each double lysogen, we can deduce the gene order therein, and from that order we can reconstruct the integration event. We do not necessarily expect each integration to be a single crossover, but we do expect the principle of progressive rarity to hold. As the gene order is fixed from the segregation data, this is a restrictive condition that tests the validity of our primary assumptions.

For simplicity, consider a case where the markers are equally spaced, so that any single crossover is more common than any triple. Four major classes are then expected: $abcABC, abCABc, aBCAbc, ABCabc$.

We expect the ratios of these classes to depend on conditions of superinfection. In a lysogenic bacterium superinfected with a homoimmune phage, integrase formation is repressed (Chap. 6). Consequently, much of the integration that does occur should be interstitial, within the prophage. On the other hand, in heteroimmune superinfection, or in superinfection of a bacterium carrying a cryptic prophage, most of the integration will be due to integrase action, which will locate the second prophage terminal to the first. So the rare doubles resulting from homoimmune superinfection should belong to all four classes *abcABC, abCABc, aBCAbc, ABCabc*; whereas heteroimmune superinfection will yield only *ABCabc* and *abcABC*.

The most extensive documentation is available for homoimmune superinfection. Calef *et al.* (1965, 1968) have shown not only that the four expected classes occur, but that their relative frequencies are in accordance with the genetic map distances known from phage crosses. Furthermore, the prophage gene order deducible from segregation analysis agrees with that already postulated on other grounds (Chap. 5). Superinfection of cells lysogenic for the cryptic prophage produces mainly terminal rather than interstitial integration, in agreement with expectation (Fischer-Fantuzzi, 1967).

The theory predicts that, regardless of the disposition of alleles, segregation analysis of each diploid should indicate the same order of gene loci. This is a further test of the adequacy of the basic assumptions made.

Prophage mapping from segregation data was first done with heterogenotes produced by superinfecting lambda lysogens with lambda *dg* (Campbell, 1963a). In this system, it is not convenient to examine all monolysogenic segregants, but only those that have lost lambda *dg* (thereby becoming gal^-), while retaining normal lambda. The other segregants (that have kept lambda *dg* but lost normal lambda) rarely, if ever, occur.

It is impossible to deduce a unique gene order for the double lysogen from such partial segregation data, employing the qualitative mapping procedure described above. Fortunately, in the case of lambda *dg* two accessory facts are helpful:

1. The fact that segregants (either gal^+ or gal^-) lacking normal lambda are very rare can be introduced into our argument. The gal^+ gene behaves formally as an allele to the phage gene block missing from the lambda *dg*. This block acts, according to our previous argument, as though very close to the limit of homology. The order must then be $\cdots gal \cdots (J)$, where (J) is one of the lambda cistrons always absent from lambda *dg*.

2. Gal^- segregants can arise not only from crossing over between the two prophages, but also between the *gal* region of the prophage and the *gal* region of the bacterial chromosome adjacent to the prophage. By using a strain with two markers in the *gal* region whose orientation on the bacterial chromosome is known, we can examine segregants recombinant between the two markers. We take the strain

galK-2 galT-1 N R A $\cdots$ *galK-2$^+$ galT-1$^+$ N R A* $\cdots$ *(J) bio*

and examine *galK-2 galT-1$^+$* segregants. Only if the two *gal* regions are oriented in the same direction (as expected from Chap. 8) will this kind of segregant generally be monolysogenic and haploid for gal. The phage genes most commonly found among these segregants are those that occur between gal and (J), and their alleles must lie between the two gal regions.

The rest of the analysis then proceeds, as above [using $f(A \cdot \cdot J)$, $f(\cdot b \cdot J)$, etc. in place of $f(A \cdot \cdot)$, $f(\cdot b \cdot)$]. The resulting prophage gene order is that predicted from Fig. 5-1. In three such lysogens examined, it happened in all cases that the complete lambda *dg* had integrated between *gal* and lambda.

PHAGE PRODUCTION FROM DOUBLE LYSOGENS

When a double lysogen produces phage, a mixture of all possible recombinant classes is obtained. Their frequencies should depend on the arrangement of markers in the lysogen. Prophage excision from a double lysogen might in principle take place by the action either of integrase or of a generalized recombinase. The former would generate a preponderance, in the lysate, of the two terminal sets of prophage alleles: A lysogen with gene arrangement *ABcabC*, for example, would produce mostly phages of types ABc and abC. A generalized recombinase would produce the four types ABc, aBc, abc and abC, any one of which might be the majority class.

Qualitatively, the fact that excision can be effected by a generalized recombinase is indicated from studies on lysogens of lambda *int*. Single lysogens of lambda *int* produce little phage on induction. However, a cell doubly lysogenic for two *int* prophages yields an approximately normal burst of *int* phage. One mode of excision open to the double and not to the single is through generalized recombinase action, as diagrammed in Fig. 12-1.

Figure 12-1 diagrams prophage loss from a growing bacterial cell. In a cell that produces phage, we would instead save the excised ring and discard what remains in the chromosome. Calef (1968) has in fact, observed that the composition of the phage produced by spontaneous lysis of double lysogens is the exact inverse of that of monolysogenic segregants. This shows that generalized recombination can predominate, even in cells where we might expect integrase to be derepressed.

In Chap. 11 we postulated that lambda can form a specific "nicking enzyme" (Fig. 11-2). This enzyme could also effect prophage excision from a double lysogen, and seems to do so under some circumstances (Signer, 1968).

ADEQUACY OF THE THEORY

The approach of the present chapter has been largely didactic. A theory of double lysogeny has been developed, and the results of experiments utilizing this theory have been presented. The demonstrated utility suggests that the theory is correct as far as it goes. Whether it is adequate is another matter.

An adequate theory of prophage attachment in general and polylysogeny in particular must explain all observed phenomenology. Specifically, it must predict the production by a given experimental procedure of as many different types of lysogens as are in fact found.

For example, lysogenization with two prophages differing by one genetic marker should produce only two types of double lysogens *Aa* and *Aa*, where the left-right orientation is with respect to a fixed vector on the bacterial chromosome. These two types should have symmetrical properties with respect to segregation pattern, phage yield after induction, etc. Similarly, with two different markers, four types are possible, *ABab, AbaB, aBAb, abAB,* and symmetry is again required.

If we look at all available data on polylysogeny, especially those collected in the pristine era before any model was available to influence interpretation, there seem frequently to have been more observed types than these considerations indicate. To give a single example, at one time (Campbell, 1963*b*), I isolated two double lysogens carrying as prophages lambda and lambda *imm434*. These differed in the rate at which monolysogenic derivatives were produced. One was relatively stable. The other was so unstable that in a single culture passage about 90 percent of the cells had lost one or the other prophage, while the remaining 10 percent perpetuated the unstable double lysogeny to the next generation.

Some results of this type undoubtedly are due to polylysogeny of order higher than two. At least three lambda prophages can be carried by one cell at one time. We have little knowledge of the frequency of multiple lysogens or their segregation properties. Until this is available, it is hard to assess the adequacy of the theory.

It seems unlikely that the theory will be disproven by these means, but, as we have seen in earlier chapters, introduction of additional elements may be necessary for comprehensive understanding.

13

TOWARD A DEFINITION OF THE EPISOME

Some problems of nomenclature arise in the distinction between episomes and plasmids, or between plasmids and chromosomes. In bacteria, where we are dealing almost entirely with genetic rather than cytogenetic facts, it is arbitrary to speak of a cell as having one chromosome and several plasmids rather than to describe it as having several chromosomes, one of which is much larger than the others. We have also mentioned that episomes can pick up genes from the chromosomes so that a cell harboring such an episome may be diploid for those genes.

Presumably, if any element containing genes represented on the host chromosome is introduced into that host, recombination between the two will occur at some frequency or other. Recombination will cause integration, and the element will thus constitute an episome. At another level, we might even contend that an element containing no chromosomal genes will nonetheless have *some* chance of becoming attached, through random breakage and joining. This is a far cry from an element such as the lambda phage, which apparently goes to the trouble of elaborating an enzyme for the specific purpose of facilitating its insertion into and excision from the bacterial chromosome.

Definitions, to be useful for the experimentalist, must be operational. But biological definitions, to be meaningful, must ultimately take cognizance of the unifying principle of natural selection for function. Lwoff's (1957) definition of a virus as something having, *inter alia,* an "organized infectious phase" is a good example. Such an entity as a transforming principle, which satisfies most of the operational definition of a virus, is excluded. Even though it can be infectious in the laboratory, and might even be so in nature, there is no indication that the cell expends any energy or elaborates any specific system that improves infectivity. The infectious phase is not *organized*.

Such a definition creates difficulties for the logical purist. It can always be argued that perhaps the DNA molecules of transformable species have evolved in such a way that they are more readily liberated, less easily degraded, or more rapidly imbibed by other cells than a random sequence of nucleotides might be. And we cannot prove that even the most complex virus is built as it is *because* this structure is well adapted to protecting nucleic acid and introducing it into other cells. Unless the whole evolutionary history of the object under study is known, naturalistic criteria cannot be rigorously applied in its classification. But the absence of logical rigor must not obscure the fact that biological objects are products of evolution, and that a classification that ignores design for function is in the last analysis arbitrary and trivial.

This being said, the only objects we want to classify as episomes ultimately, when complete knowledge of their habits is available, are those whose ability to go in and out of chromosomes is more than an accident, but which show signs of having evolved this ability. The lambda phage, which apparently elaborates an enzyme specifically catalyzing its own insertion and excision, clearly qualifies. Even though we cannot specify precisely why ability to lysogenize might be of selective value, the property is clearly no accident but rather the product of extensive evolution.

To fit transfer agents such as F, RTF, colicinogeny determinants, etc., into this picture, it helps to compare the category "transfer agents" to the category "phages," not just to "temperate phages." Some phages are temperate, but many are not. Sometimes this is for minor reasons, as in laboratory mutants unable to repress their own synthesis, or unable to insert themselves into the chromosome. Other species seem completely committed to virulence.

We could hardly claim to understand why this is so. As in other cases of applied Darwinism, we can postulate that different viruses inhabit different ecological niches. In some of these, selection is for virulence; in others, for peaceful coexistence with the host. Alternatively, we can imagine that the virulent reaction is always an accident, occurring only when the virus encounters an unnatural niche such as a foreign host.

The primary properties of a virus are replication at the expense of the host and ability to pass from one host to another. These properties serve the good of the virus and, in general, the detriment of the host. One can argue that secondary selection will ultimately create a balance where the virus does the host minimal harm or even indirectly benefits it.

All the same general considerations of the host parasite relationship apply equally to transfer agents. Their primary property is the ability to synthesize pili, which become conjugation bridges, and through which the DNA of the transfer factor can pass, replicating as it goes. In general, this serves the good of the transfer agent and the detriment of the host, insofar as energy and materials are diverted from host cell functions to synthesis of the agent and its gene products.

If we look at many transfer agents isolated from nature, we may expect to find all possible variations of coadaptation with the host. As with the phage, some agents may lack the ability to attach to the chromosome or to repress their own gene functions. We may be seeing the agents either in their natural hosts or in strains that have recently acquired them by chance infection.

For many transfer agents, no integrated state has beeen demonstrated. They qualify, technically and operationally, as plasmids, not as episomes. Those that do integrate show no evidence for a specific insertion *system* like that produced by lambda. Is the occasional ability to integrate an accident, or rather the product of natural selection?

I have no strong opinion on this question. One can reasonably argue that the survival value of transfer factors ultimately depends on their ability to facilitate host recombination. However, host chromosome transfer does not require attachment (see Chap. 4). Whether some additional advantages of attachment might be worth the trouble of bringing attachment about is unclear. It is interesting that F, the most efficient agent known for promoting chromosome transfer, can attach to many parts of the chromosome. This seems to imply at least the energy expenditure necessary for reproducing some regions of homology with the host. Whether the agent may also itself have an enzymic system specifically promoting rare, low homology recombination is an open question.

One reason experimental scientists avoid "naturalistic" or evolutionary arguments such as those presented in this chapter is their subtlety, and the danger of semantic traps. We have made a case that phage lambda, for instance, is a "true" episome because of the highly evolved system that inserts and excises prophage. The only noteworthy qualification is that this argument applies, not really to the phage as a whole, but to that section of its genome specifically concerned with prophage integration. It is possible that this section evolved as part of the bacterial chromosome, its "true" function being to cause specific breaking and joining of that chromosome at controlled times. Its association with a phage genome might be a relatively recent and evolutionarily insignificant occurrence. Certainly the heterogeneity in base composition of the different parts of the lambda genome suggests a polyphyletic origin. The possible relationships of episomes to chromosomal mechanics in general will be discussed in Chap. 14.

14

EPISOMES AS MODEL SYSTEMS

EPISOMES AND DEVELOPMENT

At the time the episome concept was formulated, one of its exciting possibilities was that it might help in understanding cellular differentiation. If we consider a lysogenic bacterial culture, for example, the cells are producing a collection of enzymes and cell constituents which collectively result in a process of harmonious balanced growth. Disturbance of this balance by rather nonspecific inducing agents causes a radical alteration in the path of synthesis—culminating in lysis of the cells and liberation of phage particles. It seemed reasonable that activation of other episomes might not be lethal but might instead induce a permanent alteration in type of synthesis, leading either to a new pattern of balanced growth or to genesis of new cell types where growth would cease.

This is still a reasonable possibility, but the problem can now be viewed in wider perspective. The feature of prophage induction that helped explain differentiation has now been fitted into a unified concept of gene repression which encompasses not only phage genes but bacterial genes as well. The ability of the prophage to go on and off the chromosome plays no essential role in this concept. Most models of differentiation that invoke episomes can be framed almost as well in the context of inducible and repressible genes that are permanently chromosomal.

One bacterial process that bears some resemblance to differentiation in higher forms is sporulation. A population of bacteria that has recently been in balanced growth will, in response to environmental conditions, so alter its pattern of synthesis that radical chemical and structural changes ensue, culminating in reduction of most of the cells from their usual form to relatively inert, highly resistant spores. It was suggested by Jacob, Schaeffer, and Wollman that activation of an episome might play a causative role in sporogenesis.

The evidence at the time (1960) was scanty. Since then, much work has been done on genetic control of sporulation in *Bacillus subtilis* (Schaeffer et al., 1965; Takahashi, 1965). The results show that many chromosomal genes are necessary for sporulation. Each gene affects a defined stage of the process.

The original suggestive evidence for episomal control of sporulation is thus no longer of weight. The observations have all been given a better explanation following more thorough analysis. In the meantime, however, new experiments, some of them motivated by the episome concept, again have provided suggestive evidence for nonchromosomal elements affecting sporulation.

Nonreverting mutants deficient in sporulation can be induced by acridine dyes, a treatment known to cure some autonomous episomes and plasmids (Rogolsky and Slepecky, 1964). On detailed analysis (Bott and Davidoff-Abelson, 1966), these mutants prove to be poor sporulators rather than nonsporulators, and to have altered respiratory abilities. This suggests that the primary genetic effect is on metabolic processes not specifically concerned with sporulation, whether or not an episome may be involved.

In multicellular eukaryotes, where real differentiation occurs, there are no demonstrated examples of true episomes. Several systems of relevance to the episome concept are known, however.

The most intensively studied of these are the "controlling elements" in maize, whose discovery stems from the pioneering work of McClintock. These elements resemble episomes in that they can add to preexisting chromosomes and thenceforth behave as normal parts thereof. Once it has added to a chromosome, a controlling element maps at a definite site. Controlling elements are occasionally lost from the chromosome (usually by transposition to some other part of the genome), and they can cause chromosome breaks at their site of attachment. No autonomous phase is known. There is one recorded case where an element can be transposed to a new location without being lost from the original position; so the element sometimes can replicate more often than the chromosome (Greenblatt and Brink, 1962).

The name *controlling element* derives from two facts: (1) Insertion of such elements frequently has a regulatory effect on activity of known genes near the site of insertion. (2) Whether or not a given element will cause chromosome breaks, undergo transposition, or regulate adjacent genes can be contingent on the presence in the genome of a second element. In the classical case first studied by McClintock, the dissociation element (*Ds*) functions only when the activator (*Ac*) is present in the complement. *Ac*, like *Ds*, is transposable and can occupy, at different times, various locations on different chromosomes.

It has been suggested that the controlling elements are heterochromatin, but the evidence is quite indirect. One of the earliest cases that in retrospect turned out to involve a controlling element was the mutator effect of the "dotted" gene in maize. This gene, located in a heterochromatic region, causes frequent mutations of certain alleles of the *A* locus, which determines aleurone color—apparently because of the presence, near the *A* locus, of an element similar to *Ds*, over which the dotted locus has a controlling effect. In *Drosophila*, translocations of heterochromatin next to euchromatin can cause variegated position effects on the euchromatic genes. This is similar to the effect of a controlling element. Unlike those changes caused by controlling elements, the changes induced in this case are not transmissible through the germ line, however.

The mode of attachment of controlling elements to chromosomes is unknown. It is not yet feasible to study the internal structure of the element itself as we do with phage. Without such information, the element will necessarily map as a point, and insertion is indistinguishable from lateral attachment. The observed chromosomal breakage at the site of attachment fits the idea of insertion (cf. Chap. 11).

At the outset of this chapter, we noted that episomes had been introduced into the discussion of differentiation by virtue of their regulatory effects; and that, according to current thinking, regulation was adequately accounted for by elements with none of the distinctive properties of episomes. Attempts have been made to consider the regulatory properties of controlling elements independently of transposability. Analogies can be drawn between the two elements *Ds* and *Ac* and the operator and regulator loci, respectively.

That regulation and transposability are attributes of the same elements seems more than coincidental, however. Not only the function of the "operator" locus *Ds* but also its ability to undergo transposition and cause chromosome breaks are strictly controlled by the presence and gene dosage of the "regulator" *Ac*. Elements like *Ac* can cause, not only reversible modifications of gene function at loci close to the insertion site of *Ds*, but also permanent, germinally transmissible mutations that map at or near the loci in question. If the *lac* repressor, for instance, can cause such a variety of effects on the *lac* operator, these have not been observed. Among the episomes however, we have seen properties adequate to account for all these effects; and, though their mechanisms are far from understood, we are at least at the stage of posing precise questions concerning them. If ordinary operator loci can show such a variety of responses, their size and complexity must be greater than currently imagined, and must approach those of known episomes.

Controlling elements clearly have considerable similarities to known bacterial episomes. Some cases of unstable inheritance in *Salmonella* can be interpreted in terms of elements even more similar to the maize factors—although their rather evanescent character makes a complete description difficult (Dawson and Smith-Keary, 1963).

TUMOR VIRUSES

One reason for interest in episomes and lysogeny has been that they might constitute a good model system for latent viruses, especially tumor viruses. The main question is not so much whether the observations on bacterial episomes can be generalized to include animal viruses, but rather what form the generalization should take. It would be unwise to anticipate too strict a parallelism between similar but distantly related phenomena.

Study of tumor viruses is more difficult than that of bacterial viruses for several reasons, especially the rudimentary state of metazoan cell genetics. The central fact of tumor virology is that exposure to certain viruses can cause a permanent, heritable alteration at the cellular level. This alteration is manifested by the uncontrolled growth characteristic of cancerous tissue. The main accomplishment of the last few years has been the demonstration that the virus genome, or a portion thereof, commonly persists in virus-induced tumors. This point had not been clear earlier for various reasons.

In 1911, Rous showed that certain tumors of birds could be caused by an infectious, self-reproducing agent. This agent is an RNA virus. The Rous sarcoma virus (RSV) itself is defective (Hanafusa *et al.*, 1964). Infectious particles are formed only in cells infected with another virus (avian leukosis virus). Some of these particles are identifiable as RSV rather than avian leukosis virus by their ability to elicit tumors. Their antigenic specificity, however, depends on that of the helper virus, indicating that the viral envelope is formed of material coded by the helper rather than by RSV.

Tumors induced by RSV frequently liberate no infectious virus. This gave credence to the notion that the relation of virus to tumor could be that of an initiating agent, inducing a potentiality latent in the host cell, and that viral information was not necessary for maintenance of the neoplastic state. This notion could still be correct, but is not required to explain the actual situation with RSV. In fact, RSV information is perpetuated in the tumor cell line and apparently can even be transmitted on occasion to neighboring cells (Vogt, 1967), although no infectious particles are detectable. The presence of RSV in the tumor cells is easily shown by infection with helper virus of tumor cell lines many generations removed from the primary infection. The fact that RSV is liberated by such infection implies that its genetic information has multiplied extensively in a latent form.

Similarly, mammalian DNA viruses such as polyoma and SV40 cause the formation of tumors which are notable for the continued presence, in tumor cell cultures, of proteins that also appear in the virus-infected cell soon after infection but are otherwise unknown in the host (Sabin and Koch, 1964). Whereas direct viral determination of protein structure has not yet been shown, the simplest interpretation is that these proteins are indeed coded by the virus genome. This implies that virus genes are perpetuated in tumor cell lines, even though infectious particles are not found and the amount of virus-specific DNA present is immeasurably small.

A sufficient explanation for viral induction of tumors would then be that the neoplastic character of the cell depends directly on the presence of virus-specific protein coded by the latent viral genome. Secondary effects on the course of differentiation and evolution of the new cell type are of course expected, as the selective pressures on the cell genome must be severely modified by the new growth habit. At any rate, we are concerned here with the manner in which the viral genes are multiplied intracellularly and distributed to both daughters at cell division in the apparent absence of external reinfection.

Lysogeny provides a simple model. The basic features of lysogeny are self-control (through immunity repression) and integration into the chromosome. All elements able to perpetuate themselves indefinitely in bacterial cells display the first property. Those that also show the second are termed episomes. Most overt manifestations of lysogeny are consequences of the first property. The second is detectable only by careful genetic analysis.

In bacteria, then, some extrachromosomal elements attach to the chromosome, whereas others, also capable of indefinite propagation in a living cell line, are autonomous. The model for polyoma could be RTF as easily as it might be lambda. The question of chromosomal attachment can be decided only by direct experiments on the tumor viruses, which hopefully soon will be feasible.

Regardless of the outcome of such experiments, one fact is clear:

Like bacteria, mammalian cells can acquire genetic information by infection, and can perpetuate such information indefinitely and intracellularly. This fact must not be neglected in any formulation of the evolution of these cells and of the organisms they comprise.

EPISOMES AND CHROMOSOMAL MECHANICS

The quest for the physical basis of biological specificity has led, in the present generation, to the identification of DNA base sequences as the ultimate library in which information is stored and reproduced. The secret of life does not reside in DNA, however. The high degree of specificity, precisely replicated, is irrelevant except in the context of a living cell, where an alteration of sequence is reflected in differential expression.

Differential expression requires specific recognition by the cell of configurations in the DNA molecule. The most familiar and general example is the translation of nucleic acid sequence into protein sequence. It is an elaborate process involving many different elements, which ultimately effect recognition of, and discrimination between, different base triplets. The code is universal, except for minor modifications. In the presence of this translational system, each DNA molecule expresses some of its individuality by the proteins for which it codes.

Such universal systems are just the beginning of an understanding of specificity, however. Besides these general mechanisms, common to all cells, particular cells have specific, gene-controlled systems that recognize rare configurations found at unique spots in the cell genome and control genetic activity at the site of these configurations. This was shown to be true for gene function by Jacob and Monod.

This is again likely to be just the beginning. Not only the function of genes, but their assortment at cell division, their position along the chromosomes, etc., are probably determined by specific recognition mechanisms.

Molecular biologists have concentrated attention on bacteria and their viruses because of their technical advantages for controlled experimentation. The faith that principles and mechanisms discovered with these lower forms might be of general applicability to other living things has been amply justified during the last decade. We can be confident that many of the tricks known to bacterial episomes are employed at other places in different contexts.

This conclusion applies more to individual mechanistic steps than to overall processes. We have seen that episomes can elaborate specific systems for catalyzing their own replication and transcription, and also for cutting and joining very specific genetic regions, precisely in some cases and imprecisely in others. Even in so simple a creature as the lambda phage, these processes are under a refined system of genetic regulation that subordinates the individual steps to the master plan of an integrated life cycle.

I would be surprised if these same mechanisms were not employed, in an equally purposeful manner, in the life cycle of organisms much more complex and highly evolved. This need not imply that individual chromosome segments are continually popping off the chromosome, multiplying in the cytoplasm, and reintegrating at later stages of development. The most fruitful analogies may not be at nearly so gross a level.

In the previous chapter, a case was made that the lambda phage is a "true" episome in the sense that its chromosomal attachment was far from accidental but depended on a highly evolved mechanism. We mentioned also the alternative that the "chromosomal attachment system" was in fact derived by minor modifications of a system concerned with controlled, specific breakage and joining of the host chromosome. We did not justify the expectation that the host might have or need such a system; and indeed, bacteria provide little information on this point. If we consider all the observations on chromosome behavior in higher organisms, however, there is a general indication that highly specific systems must play a large role in controlling not only gene function but also chromosomal mechanics. Nonspecific systems such as generalized recombinases, polymerases, etc., do not seem adequate to explain what one sees. These should be viewed together with the classical embryological problem of irreversible loss of cellular potentialities during development.

I therefore list here, without detailed comment, some of the most pertinent observations: At specific stages of development, under control of hormonal stimuli, specific chromosomal segments form "puffs" —areas rich in protein and RNA, apparently sites where genes are functioning very rapidly. There is evidence that many eukaryotic chromosomes (and even some prokaryotic ones) have extensive tandem duplications of genetic material. It has been suggested (Callan, 1967) that some of this duplicated DNA, although inserted into the continuity of the chromosome, is not replicated as such, but rather that a whole series of duplicated regions are copied from one terminal "master region" at each cell division.

A wide variety of amino acid sequences are found among samples of the human Kappa light immunoglobulin chain. Yet genetic studies indicate that only one gene codes for that portion of the sequence common to this family of molecular species. It is now clear, also, that the antibody-forming cells of a single organism are differentiated into a large number of specific clones, each with strongly restricted potentialities. These facts have led to the proposal of an orderly program by which a single gene becomes successively "spliced onto" other genetic regions during the course of differentiation (Dreyer *et al.,* 1967).

Whether or not these particular suggestions turn out to be correct, it seems likely that the complex, orderly behavior displayed by lampbrush chromosomes, by chromosome puffs, etc., will find its explanation in the presence of enzymes that recognize specific nucleotide sequences and act upon them. Study of bacterial episomes seems to be leading us toward an understanding of such enzymes, how they function, and what potentialities they may have. This is likely to be the main contribution of the episome concept to biology.

REFERENCES

The following is a list of references cited in the text. Listing of a reference does not imply that the paper is worth reading. I have tried primarily to document statements made in the text. My experience in using this sort of book in teaching is that I frequently encounter statements framed by the author on the basis of concepts current in his field at the time of writing, and for which I am unable to reconstruct the factual basis. With a few exceptions, no references before 1961 are included. These were covered in my review article (*Advances in Genetics,* **11:**101, 1962), which should be accessible in libraries containing the other journals mentioned.

During the time of writing this book (1966–1968), several review articles and one book (all highly recommended) have also been prepared. These are listed separately. I have tried to modify the book to include some new ideas and facts from these sources, but it has been impossible to rewrite each chapter completely to accommodate new material. I am indebted to Drs. Dove, Echols, Falkow, and Signer for copies of their manuscripts prior to publication.

ABE, A., and J. TOMIZAWA. 1967. Replication of the *Escherichia coli* K-12 chromosome. *Proc. Natl. Acad. Sci. U.S.,* **58:** 1911–1918.

ADELBERG, EDWARD A., and JAMES PITTARD. 1965. Chromosome transfer in bacterial conjugation. *Bacteriol. Rev.,* **29:** 161–172.

ADHYA, S. 1968. Personal communication.

ADHYA, S., P. P. CLEARY, and A. CAMPBELL. 1968. Nitrate reductase mutants of *Escherichia coli. Bacteriol. Proc.,* 55.

ANDERSON, E. S., and M. J. LEWIS. 1965. Characterization of a transfer factor associated with drug resistance in *Salmonella typhimurium. Nature,* **208:** 843–849.

BALDWIN, R. L., P. BARRAND, A. FRITSCH, D. A. GOLDTHWAIT, and F. JACOB. 1966. Cohesive sites on the deoxyribonucleic acids from several temperate colophages. *J. Mol. Biol.*, **17**: 343–357.

BECKWITH, J. R., E. R. SIGNER, and W. EPSTEIN. 1966. Transposition of the *lac* region of *E. coli*. *Cold Spring Harbor Symposia Quant. Biol.*, **31**: 393–402.

BENZER, S. 1959. On the topology of the genetic fine structure. *Proc. Natl. Acad. Sci. U.S.*, **45**: 1607–1620.

BERG, C. M., and L. G. CARO. 1967. Chromosome replication in *Escherichia coli*. I. Lack of influence of the integrated F factor. *J. Mol. Biol.*, **29**: 419–432.

BERG, C., and ROY CURTISS, III. 1967. Transposition derivatives of an Hfr strain of *Escherichia coli* K-12. *Genetics*, **56**: 503–525.

BERTANI, G. 1962. Multiple lysogency from single infection. *Virology*, **18**: 131–139.

BERTANI, L. E. 1965. Limited multiplication of phages superinfecting lysogenic bacteria and its implication for the mechanism of immunity. *Virology*, **27**: 496–511.

BODE, VERNON C., and A. D. KAISER. 1965. Repression of the *cII* and *cIII* cistrons of phage lambda in a lysogenic bacterium. *Virology*, **25**: 111–121.

BONHOEFFER, FRIEDRICH. 1966. DNA transfer and DNA synthesis during bacterial conjugation. *Z. Vererbungsl.*, **98**: 141–149.

BOTT, K. F., and R. DAVIDOFF-ABELSON. 1966. Altered sporulation and respiratory patterns in mutants of *Bacillus subtilis* induced by acridine orange. *J. Bacteriol.*, **92**: 229–240.

BOTT, K., and B. STRAUSS. 1965. The carrier state of *Bacillus subtilis* infected with the transducing bacteriophage SP10. *Virology*, **25**: 212–225.

BRINTON, CHARLES C. 1965. Structure, function, synthesis and genetic control of bacterial pili and a molecular model for DNA and RNA transport in gram negative bacteria. *Trans. N.Y. Acad. Sci.*, **27**: 1003–1054.

BRONSON, M., and B. L. KELLY. 1967. Temperature-sensitive mutants of bacteriophage P2. *Bacteriol. Proc.*, V121.

BROOKS, KATHERINE. 1965. Studies in the physiological genetics of some suppressor-sensitive mutants of bacteriophage lambda. *Virology*, **26**: 489–499.

BROOKS, KATHERINE, and A. CLARK. 1967. Behavior of λ bacteriophage in a recombination deficient strain of *Escherichia coli*. *J. Virology*, **1**: 283–293.

CALEF, E. 1968. Mapping of integration and excision crossovers in superinfection double lysogens for phage lambda in *Escherichia coli*. *Virology* (in press).

CALEF, E., C. MARCHELLI, and F. GUERRINI. 1965. The formation of superinfection double lysogens of phage in *Escherichia coli* K-12. *Virology*, **27**: 1–10.

REFERENCES

CALENDAR, R., and G. LINDAHL. 1968. Personal communication.

CALLAN, H. G. 1967. The organization of genetic units into chromosomes. *J. Cell. Sci.*, **2**: 1–7.

CAMPBELL, ALLAN. 1963*a*. Segregants from lysogenic heterogenotes carrying recombinant lambda prophages. *Virology*, **20**: 344–356.

CAMPBELL, ALLAN. 1963*b*. Distribution of genetic types of transducing lambda phages. *Genetics*, **48**: 409–421.

CAMPBELL, ALLAN. 1964. Genetic recombination between λ prophage and irradiated λ*dg* phage. *Virology*, **23**: 234–251.

CAMPBELL, ALLAN. 1965*a*. The steric effect in lysogenization by bacteriophage lambda. I. Lysogenization of a partially diploid strain of *Escherichia coli* K-12. *Virology*, **27**: 329–339.

CAMPBELL, ALLAN. 1965*b*. The steric effect in lysogenization by bacteriophage lambda. II. Chromosomal attachment of the *b2* mutant. *Virology*, **27**: 340–345.

CAMPBELL, ALLAN. 1967*a*. In vitro reconstitution of lambda *dg*. *Bacteriol. Proc.*, 155.

CAMPBELL, ALLAN. 1967*b*. Regulation in viruses. In *Molecular Genetics*, Vol. 2. J. H. Taylor, ed. Academic Press, New York, pp. 323–382.

CAMPBELL, ALLAN. 1968. Metric and topological approaches to genetic mapping. In *Philosophy of Science, the Delaware Series*, Vol. 3. Wiley, New York (in press).

CAMPBELL, ALLAN, and ALICE DEL CAMPILLO CAMPBELL. 1963. Mutant of lambda bacteriophage producing a thermolabile endolysin. *J. Bacteriol.*, **85**: 1202–1207.

CAMPBELL, ALLAN, and KAREN KILLEN. 1967. Effect of temperature on prophage attachment and detachment during heteroimmune superinfection. *Virology*, **33**: 749–752.

CAMPBELL, ALLAN, and JAMES ZISSLER. 1966. The steric effect in lysogenization by bacteriophage lambda. III. Superinfection of monolysogenic derivatives of a strain diploid for the prophage attachment site. *Virology*, **28**: 659–662.

DEL CAMPILLO CAMPBELL, A. 1968. Personal communication.

DEL CAMPILLO CAMPBELL, ALICE, GARY KAYAJANIAN, ALLAN CAMPBELL, and SANKAR ADHYA, 1967. Biotin-requiring mutants of *Escherichia coli* K-12. *J. Bacteriol.*, **94**: 2065–2066.

CARO, LUCIEN G. 1965. The molecular weight of lambda DNA. *Virology*, **25**: 226–236.

CAIRNS, JOHN. 1963. The chromosome of *Escherichia coli*. *Cold Spring Harbor Symposia Quant. Biol.*, **28**: 43–46.

CHAI, N., and K. G. LARK. 1967. Segregation of deoxyribonucleic acid in bacteria: association of the segregating unit with the cell envelope. *J. Bacteriol.*, **94**: 415–421.

CLOWES, R. C., and E. E. M. MOODY. 1966. Chromosomal transfer from "recombination-deficient" strains of *Escherichia coli* K-12. *Genetics*, **53**: 717–726.

CLOWES, R. C., E. E. M. MOODY, and R. H. PRITCHARD. 1965. The elimination of extrachromosomal elements in thymineless strains of *Escherichia coli* K-12. *Genet. Res. Camb.*, **6:** 147–152.

CURTISS, ROY, III. 1964a. An *Escherichia coli* K-12 Hfr strain attached to or within the structural gene for β-galactosidase. *Bacteriol. Proc.*, 30.

CURTISS, ROY, III. 1964b. A stable partial diploid strain of *Escherichia coli*. *Genetics*, **50:** 679–694.

CURTISS, ROY, III, and JANET RENSHAW. 1968. Chromosome mobilization and transfer in $F^+ \times F^-$ matings in *Escherichia coli*. I. Evidence for chromosome transfer in the absence of F integration. *Genetics* (in press).

CUZIN, FRANCOIS. 1962. Multiplication autonome de l'episome sexuel d'*Escherichia coli* K12, dans une souche Hfr. *Compt. rend.*, **254:** 4211–4213.

CUZIN, F., and F. JACOB. 1965. Analyse genetique fonctionelle de l'episome sexuel d'*Escherichia coli* K12. *Compt. rend.*, **260:** 2087–2090.

CUZIN, FRANCOIS, and F. JACOB. 1967a. Existence chez *Escherichia coli* K12 d'une unite genetique de transmission formee de differents replicons. *Ann. inst. Pasteur*, **112:** 529–545.

CUZIN, FRANCOIS, and F. JACOB. 1967b. Mutations de l'episome F d'*Escherichia coli* K12. II. Mutants a replication thermosensible. *Compt. rend.*, **112:** 1–9.

DATTA, NAOMI. 1965. Infectious drug resistance. *British Med. Bull.*, **21:** 254–259.

DATTA, NAOMI, and M. H. RICHMOND. 1966. The purification and properties of a penicillinase whose synthesis is mediated by an R factor in *Escherichia coli*. *Biochem. J.*, **98:** 204–209.

DAWSON, G. W. P., and P. F. SMITH-KEARY. 1963. Episomic control of mutation in *Salmonella typhimurium*. *Heredity*, **18:** 1–20.

DOVE, W. F. 1966. Action of the lambda chromosome. I. Control of functions late in bacteriophage development. *J. Mol. Biol.*, **19:** 187–201.

DOVE, W. F. 1967. The synthesis of the λ chromosome: the role of prophage termini. In *The Molecular Biology of Viruses*. J. Colter and W. Paranchych, eds. Academic Press, New York, pp. 111–124.

DREYER, W. J., W. R. GRAY, and L. HOOD. 1967. The genetic, molecular and cellular basis of antibody formation: some facts and a unifying hypothesis. *Cold Spring Harbor Symposia Quant. Biol.*, **32:** 353–367.

DUBNAU, E., and B. A. D. STOCKER. 1964. Genetics of plasmids in *Salmonella typhimurium*. *Nature*, **204:** 1112–1113.

ECHOLS, H., R. GINGERY and L. MOORE. 1968. The integrative recombination function of bacteriophage λ—evidence for a site-specific recombination enzyme. *Proc. Natl. Acad. Sci. U.S.* (in press).

ECHOLS, H., L. PILARSKI, and P. Y. CHENG. 1968. In vitro repression of phage λDNA transcription by a partially purified repressor from lysogenic cells. *Proc. Natl. Acad. Sci. U.S.* (in press).

EISEN, H. A., C. R. FUERST, L. SIMINOVITCH, R. THOMAS, L. LAMBERT, L. PEREIRA DA SILVA, and F. JACOB. 1966. Genetics and physiology of defective lysogeny in K12 (λ): Studies of early mutants. *Virology,* **30:** 224–241.

EISEN, H. A., L. SIMINOVITCH, and P. T. MOHIDE. 1968a. Excision of lambda prophage: effects on host survival. *Virology,* **34:** 97–103.

EISEN, H. A., I. TALLAN, and L. SIMINOVITCH. 1968b. Excision of the lambda prophage: effects on linkage relationships of *Escherichia coli* chromosome. *Virology,* **34:** 104–111.

EISERLING, F. A., and E. BOY DE LA TOUR. 1965. Capsomeres and other structures observed on some bacteriophages. *Path. Microbiol.,* **28:** 175–180.

FALKOW, STANLEY, and R. V. CITARELLA. 1965. Molecular homology of F-merogenote DNA. *J. Mol. Biol.,* **12:** 138–151.

FALKOW, STANLEY, R. V. CITARELLA, J. A. WOHLHIETER, and TSUTOMU WATANABE. 1966. The molecular nature of R factors. *J. Mol. Biol.,* **17:** 102–116.

FISCHER-FANTUZZI, L. 1967. Integration of λ and λb2 genomes in nonimmune host bacteria carrying a cryptic prophage. *Virology,* **32:** 18–32.

FISCHER-FANTUZZI, L., and E. CALEF. 1964. A type of lambda prophage unable to confer immunity. *Virology,* **23:** 209–216.

FRANKLIN, N. C. 1967. Deletions and functions of the center of the φ80-phage genome. Evidence for a phage function promoting genetic recombination. *Genetics,* **57:** 301–318.

FRANKLIN, N. C., W. F. DOVE, and C. YANOFSKY. 1965. A linear insertion of a prophage into the chromosome of *E. coli* shown by deletion mapping. *Biochem. Biophys. Res. Commun.,* **18:** 910–923.

FREDERICQ, PIERRE. 1963. Linkage of colicinogenic factors with an F agent and with nutritional markers in the chromosome and in an episome of *Escherichia coli. Proc. 11th Intern. Congr. Genet., The Hague,* 42–43.

FREIFELDER, D. 1968. Personal communication.

GANESAN, A. T., and JOSHUA LEDERBERG. 1965. A cell-membrane bound fraction of bacterial DNA. *Biochem. Biophys. Res. Commun.,* **18:** 824–835.

GAREN, A., and H. ECHOLS. 1962. Genetic control of induction of alkaline phosphatase synthesis in *E. coli. Proc. Natl. Acad. Sci. U.S.,* **48:** 1398–1402.

GELLERT, M. 1967. Formation of covalent circles of lambda DNA by *E. coli* extracts. *Proc. Natl. Acad. Sci. U.S.,* **57:** 148–155.

GILBERT, WALTER, and BENNO MÜLLER-HILL. 1966. Isolation of the lac repressor. *Proc. Natl. Acad. Sci. U.S.,* **56:** 1891–1898.

GINGERY, R., and H. ECHOLS. 1967. Mutants of bacteriophage unable to integrate into the host chromosome. *Proc. Natl. Acad. Sci. U.S.,* **58:** 1507–1514.

GOTTESMAN, M. E., and M. B. YARMOLINSKY. 1968. Integration-negative mutants of bacteriophage lambda. *J. Mol. Biol.,* **31:** 487–505.

GOUGH, M., and M. LEVINE. 1968. The circularity of the phage P22 linkage map. *Genetics,* **58:** 161–169.

GRATIA, J. P. 1966. Studies on defective lysogeny due to chromosomal deletion in *Escherichia coli*. I. Single lysogens. *Biken J.,* **9:** 77–87.

GRATIA, J. P. 1967. Production de particules douees a la fois de proprietes infectieuses et transductrices par des souches d'*Escherichia coli* lysogenes defectives par suite d'une deletion chromosomique. *Life Sciences,* **2:** 209–212.

GREENBLATT, IRWIN M., and R. ALEXANDER BRINK. 1962. Twin mutations in medium variegated pericarp in maize. *Genetics,* **47:** 489–507.

GROSS, JULIAN D. 1964. Conjugation in bacteria. In *The Bacteria*. I. C. Gunsalus and R. Y. Stanier, eds., Academic Press, New York, pp. 1–48.

GROSS, JULIAN D., and LUCIEN CARO. 1966. DNA transfer in bacterial conjugation. *J. Mol. Biol.,* **16:** 269–284.

HANAFUSA, H., T. HANAFUSA, and H. RUBIN. 1964. Analysis of the defectiveness of Rous Sarcoma Virus. I. Characterization of the helper virus. *Virology,* **22:** 591–601.

HARADA, KENJI, MITSUO KAMEDA, MITSUE SUZUKI, and SUSUMU MITSUHASHI. 1963. Drug resistance of enteric bacteria. II. Transduction of transmissible drug-resistance (R) factors with phage epsilon. *J. Bacteriol.,* **86:** 1332–1338.

HARRIS, A. W., D. W. A. MOUNT, C. R. FUERST, and L. SIMINOVITCH. 1967. Mutations in bacteriophage lambda affecting host cell lysis. *Virology,* **32:** 553–569.

HARUNA, I., and S. SPIEGELMAN. 1966. Selective interference with viral RNA formation in vitro by specific inhibition with synthetic polynucleotides. *Proc. Natl. Acad. Sci. U.S.,* **56:** 1333–1338.

HASHIMOTO, HAJIME, and YUKINORI HIROTA. 1966. Gene recombination and segregation of resistance factor R in *Escherichia coli*. *J. Bacteriol.,* **91:** 51–62.

HERSHEY, A. D., and E. BURGI. 1965. Complementary structure of interacting sites at the ends of λDNA molecules. *Proc. Natl. Acad. Sci. U.S.,* **53:** 325–328.

HERSHEY, A. D., E. BURGI, and L. INGRAHAM. 1963. Cohesion of DNA molecules isolated from phage lambda. *Proc. Natl. Acad. Sci. U.S.,* **49:** 748–755.

HIROTA, YUKINORI, TOSHIO FUJII, and YUKINOBU NISHIMURA. 1966. Loss and repair of conjugal fertility and infectivity of the resistance factor and sex factor in *Escherichia coli*. *J. Bacteriol.,* **91:** 1298–1304.

HIROTA, YUKINORI, and HISAO UCHIDA. 1964. Effects of radioactive decay on the autonomous replication of an *Escherichia coli* episome. *Z Vererbungsl.,* **95:** 184–194.

HOFFMAN, DONALD B., JR., and IRWIN RUBENSTEIN. 1968. Physical studies of lysogeny (in manuscript).

HOGNESS, DAVID S., WALTER DOERFLER, J. BARRY EGAN, and LINDSAY W. BLACK. 1966. The position and orientation of genes in λ and λdg DNA. *Cold Spring Harbor Symposia Quant. Biol.,* **31:** 129–138.

HORIUCHI, T., J. TOMIZAWA, and A. NOVICK. 1962. Isolation and properties of bacteria capable of high rates of β-galactosidase synthesis. *Biochim. et. Biophys. Acta,* **55:** 152–163.

HOWARD-FLANDERS, PAUL, RICHARD P. BOYCE, and LEE THERIOT. 1966. Three loci in *Escherichia coli* K-12 that control the excision of pyrimidine dimers and certain other mutagen products from DNA. *Genetics,* **53:** 1119–1136.

HSU, YU-CHIH. 1967. Mutation in a viral DNA genome which allows perpetuation in host cells. *Bacteriol. Proc.,* V127.

IKEDA, H., and J. TOMIZAWA. 1965. Transducing fragments in generalized transduction by phage P1. I. Molecular origin of the fragments. *J. Mol. Biol.,* **14:** 85–109.

INMAN, ROSS B. 1966. A denaturation map of the phage DNA molecule determined by electron microscopy. *J. Mol. Biol.,* **18:** 464–476.

IPPEN, KARIN, JEFFREY H. MILLER, JOHN SCAIFE, and JONATHAN BECKWITH. 1968. New controlling element in the *lac* operon of *E. coli. Nature,* **217:** 825–827.

ISRAEL, J. V., THOMAS F. ANDERSON, and MYRON LEVINE. 1967. In vitro morphogenesis of phage P22 from heads and base-plate parts. *Proc. Natl. Acad. Sci. U.S.,* **57:** 284–291.

JACOB, F., S. BRENNER, and F. CUZIN. 1963. On the regulation of DNA replication in bacteria. *Cold Spring Harbor Symposia Quant. Biol.,* **28:** 329–348.

JACOB, F., C. FUERST, and E. WOLLMAN. 1957. Recherches sur les bactéries lysogènes défectives. II. Les types physiologiques liés aux mutations du prophages. *Ann. Inst. Pasteur,* **93:** 724–753.

JACOB, F., and J. MONOD. 1961. Genetic regulatory mechanisms in the synthesis of proteins. *J. Mol. Biol.,* **3:** 318–356.

JACOB, F., and E. L. WOLLMAN. 1957. Genetic aspects of lysogeny. In *The Chemical Basis of Heredity.* W. D. McElroy and B. Glass, eds. Johns Hopkins Press, Baltimore, pp. 468–498.

JACOB, F., and E. L. WOLLMAN. 1961. *Sexuality and the Genetics of Bacteria.* Academic Press, New York.

JORDAN, E. 1965. The location of the *b2* deletion of bacteriophage λ. *J. Mol. Biol.,* **10:** 341–344.

JORDAN, ELKE, and MATTHEW MESELSON. 1965. A discrepancy between genetic and physical lengths on the chromosome of bacteriophage lambda. *Genetics,* **51:** 77–86.

JOYNER, ANNE, L. N. ISAACS, H. ECHOLS, and W. SLY. 1966. DNA replication and messenger RNA production after induction of wild type lambda phage and lambda mutants. *J. Mol. Biol.,* **19:** 174–186.

KAHN, PHYLLIS L., and DONALD R. HELINSKI. 1965. Interaction between colicinogenic factor V and the integrated F factor in an Hfr strain of *Escherichia coli. J. Bacteriol.,* **90:** 1276–1282.

KAISER, A. D. 1962. The production of phage chromosome fragments and their capacity for genetic transfer. *J. Mol. Biol.,* **4:** 275–287.

KAYAJANIAN, G. 1968. Personal communication.

KAYAJANIAN, G., and A. CAMPBELL. 1966. The relationship between heritable physical and genetic properties of selected gal^- and gal^+ transducing λdg. *Virology,* **30:** 482–492.

LARK, KARL G. 1966. Regulation of chromosomal replication and segregation in bacteria. *Bacteriol. Rev.,* **30:** 3–32.

LARK, K. G., R. A. CONSIGLI, and H. MINOCHA. 1966. Segregation of sister chromatids in mammalian cells. *Science,* **154:** 1202–1204.

LEDERBERG, J., E. M. LEDERBERG, N. D. ZINDER, and E. R. LIVELY. 1951. Recombination analysis of bacterial heredity. *Cold Spring Harbor Symposia Quant. Biol.,* **16:** 413–441.

LENNOX, E. 1955. Transduction of linked genetic characters of the host by bacteriophage P1. *Virology,* **1:** 190–206.

LIEB, MARGARET. 1966. Studies of heat-inducible mutants. II. Production of *cI* product by superinfecting lambda$^+$ in heat-inducible lysogens. *Virology,* **29:** 367–376.

LINDAHL, G. 1967. Personal communication.

LOW, BROOKS. 1967. Inversion of transfer modes and sex factor-chromosome interactions in conjugation in *Escherichia coli. J. Bacteriol.,* **93:** 98–106.

LURIA, S., J. ADAMS, and R. TING. 1960. Transduction of lactose-utilizing ability among strains of *E. coli* and *S. dysenteriae* and the properties of the transducing phage particles. *Virology,* **12:** 348–390.

LWOFF, ANDRE. 1957. Bacteriophage as a model of host-virus relationship. In *The Viruses,* Vol. 2. M. Burnet and W. A. Stanley, eds. Academic Press, New York, pp. 187–201.

MAAS, RENATA. 1963. Exclusion of an Flac episome by an Hfr gene. *Proc. Natl. Acad. Sci. U.S.,* **50:** 1951–1955.

MAAS, WERNER K., and A. J. CLARK. 1964. Studies on the mechanism of repression of arginine biosynthesis in diploids. II. Dominance of repressibility in diploids. *J. Mol. Biol.,* **8:** 365–370.

MACHATTIE, L., and C. A. THOMAS. 1964. DNA from bacteriophage lambda: Molecular length and conformation. *Science,* **144:** 1142–1144.

MANDEL, M. 1967. Infectivity of Phage *P2* DNA in Presence of Helper Phage. *Molec. Gen. Genetics,* **99:** 88–96.

MARCHELLI, C., L. PICA, and A. SOLLER. 1968. The cryptogenic factor in λ. *Virology,* **34:** 650–663.

MATNEY, THOMAS S., EUGENE P. GOLDSCHMIDT, NANCY S. ERWIN, and RUTH ANN SCROOGS. 1964. A preliminary map of genomic sites for F attachment in *Escherichia coli K-12. Biochem. Biophys. Res. Commun.,* **17:** 278–281.

MESELSON, M., and J. WEIGLE. 1961. Chromosome breakage accompanying genetic recombination in bacteriophage. *Proc. Natl. Acad. Sci. U.S.,* **47:** 857–868.

MEYNELL, ELINOR, and NAOMI DATTA. 1967. Mutant drug resistant factors of high transmissibility. *Nature,* **214:** 885–887.

MITSUHASHI, SUSUMU, HAJIME HASHIMOTO, RYUKI EGAWA, TOKUMITSU TANAKA, and YUTAKA NAGAI. 1967. Drug resistance of enteric bacteria. IX. Distribution of R factors in gram-negative bacteria from clinical sources. *J. Bacteriol.,* **93:** 1242–1245.

NAGATA, TOSHIO. 1963. The molecular synchrony and sequential replication of DNA in *Escherichia coli. Proc. Natl. Acad. Sci. U.S.,* **49:** 551–559.

NAGEL DE ZWAIG, R. 1966. Association between colicinogenic and fertility factors. *Genetics,* **54:** 381–390.

NOMURA, M. 1967. Colicins and related bacteriocins. *Ann. Rev. Microbiol.,* **21:** 257–284.

NOVICK, AARON, and TADAO HORIUCHI. 1961. Hyper-production of β-galactosidase by *Escherichia coli* bacteria. *Cold Spring Harbor Symposia Quant. Biol.,* **26:** 239–245.

NOVICK, R. P. 1967. Penicillinase plasmids of *Stephylococcas aureus. Fed. Proc.,* **26:** 29–38.

OKUBO, S., STODOLSKY, K. BOTT, and B. STRAUSS. 1963. Separation of the transforming and viral deoxyribonucleic acids of a transducing bacteriophage of *Bacillus subtilis. Proc. Natl. Acad. Sci. U.S.,* **50:** 679–686.

PACKMAN, S., and W. S. SLY. 1968. Constitutive λDNA replication by λC17, a regulatory mutant related to virulence. *Virology,* **34:** 778–789.

PARKINSON, J. S. 1968. Personal communication.

PEARCE, L. E., and ELINOR MEYNELL. 1968. Specific chromosomal affinity of a resistance factor. *J. Gen. Microbiol.,* **50:** 159–172.

PEREIRA DA SILVA, L. H., and FRANCOIS JACOB. 1967. Induction of *cII* and *O* functions in early defective lambda prophages. *Virology,* **33:** 618–620.

PEREIRA DA SILVA, LUIZ, HARVEY EISEN, and FRANCOIS JACOB. 1968. Sur la replication du bacteriophage λ. *Compt. rend.,* **266:** 926–928.

PITTARD, JAMES. 1965. Effect of integrated sex factor on transduction of chromosomal genes in *Escherichia coli. J. Bacteriol.,* **89:** 680–686.

PTASHNE, MARK. 1965a. The detachment and maturation of conserved prophage DNA. *J. Mol. Biol.,* **11:** 90–96.

PTASHNE, MARK. 1965b. Replication and host modification of DNA transferred during bacterial mating. *J. Mol. Biol.,* **11:** 829–838.

PTASHNE, M. 1967a. Isolation of the λ phage repressor. *Proc. Natl. Acad. Sci. U.S.,* **57:** 306–313.

PTASHNE, MARK. 1967b. Specific binding of the λ phage repressor to λDNA. *Nature,* **214:** 232–234.

PTASHNE, MARK, and NANCY HOPKINS. 1968. The operators controlled by the λ phage repressor. *Proc. Natl. Acad. Sci. U.S.* (in press).

RADDING, CHARLES M., JOSIANE SZPIRER, and RENE THOMAS. 1967. The structural gene for λ exonuclease. *Proc. Natl. Acad. Sci. U.S.,* **57:** 277–283.

RIS, HANS, and BARBARA CHANDLER. 1963. The ultrastructure of genetic systems in prokaryotes and eukaryotes. *Cold Spring Harbor Symposia Quant. Biol.,* **28:** 1–8.

ROGOLSKY, M., and R. SLEPECKY. 1964. Elimination of a genetic determinant for sporulation of *Bacillus subtilis* with acriflavine. *Biochem. Biophys. Res. Commun.,* **16:** 204–208.

ROLFE, B., and P. CLEARY. 1968. Personal communication.

ROTHMAN, JUNE L. 1965. Transduction studies on the relation between prophage and host chromosome. *J. Mol. Biol.,* **12:** 892–912.

ROWND, R., RINTARO NAKAYA, and AKIBO NAKAMURA. 1966. Molecular nature of the drug resistance factors of the Enterobacteriaceae. *J. Mol. Biol.,* **17:** 376–393.

SABIN, A. B., and M. A. KOCH. 1964. Source of genetic information for specific complement fixing antigens in SV40 virus-induced tumors. *Proc. Natl. Acad. Sci. U.S.,* **52:** 1131–1138.

SCAIFE, JOHN. 1967. Episomes. *Ann. Rev. Microbiol.,* **21:** 601–638.

SCAIFE, JOHN, and A. P. PEKHOV. 1964. Deletion of chromosomal markers in association with F-prime factor formation in *Escherichia coli* K12. *Genet. Res. Camb.,* **5:** 495–498.

SCHAEFFER, P., H. IONESCO, A. RYTER, and G. BALASSA. 1965. La sporulation de *Bacillus subtilis* etude genetique et physiologique. Colloq. Intern. Centre Natl. Rech. Sci. Mechanismes Regulation, Marseilles, 1963, pp. 553–563.

SCOTT, J. R. 1968. Personal communication.

SHAPIRO, J. 1967. Personal communication.

SHAW, W. V. 1967. The enzymatic acetylation of chloramphenicol by extract of R factor-resistant *Escherichia coli*. *J. Biol. Chem.,* **242:** 687–693.

SHEPPARD, DAVID, and ELLIS ENGLESBERG. 1966. Positive control in the L-arabinose gene enzyme complex of *Escherichia coli* B/r as exhibited with stable merodiploids. *Cold Spring Harbor Symposia Quant. Biol.,* **31:** 345–347.

SIGNER, E. R. 1964. Recombination between coliphages λ and $\phi 80$. *Virology,* **22:** 650–651.

SIGNER, E. R. 1966. Interactions of prophages at the $att80$ site with the chromosome of *Escherichia coli*. *J. Mol. Biol.,* **15:** 243–255.

SIGNER, E. 1968. Lysogeny: the integration problem. *Ann. Rev. Microbiol.* (in press).

SIGNER, E. R., and J. R. BECKWITH. 1966. Transposition of the *lac* region of *Escherichia coli*. III. The mechanism of attachment of bacteriophage $\phi 80$ to the bacterial chromosome. *J. Mol. Biol.,* **22:** 33–51.

SIGNER, E. R., J. R. BECKWITH, and S. BRENNER. 1965. Mapping of suppressor loci in *Escherichia coli*. *J. Mol. Biol.,* **14:** 153–166.

SILVER, S., and H. OZEKI. 1962. Transfer of deoxyribonucleic acid accompanying the transmission of colicinogenic properties by cell mating. *Nature,* **195:** 873–878.

SIMINOVITCH, L. 1967. Personal communication.

REFERENCES

SIX, E. 1961. Inheritance of prophage P2 in superinfection experiments. *Virology*, **14:** 220–233.

SIX, ERICH. 1966. Specificity of P2 for prophage site I on the chromosome of *Escherichia coli* strain C. *Virology*, **29:** 106–125.

SIX, E. W. 1968. Prophage site specificities of P2 phages. *Bacteriol. Proc.*, 159.

SKALKA, A. 1966. Regional and temporal control of genetic transcription in phage lambda. *Proc. Natl. Acad. Sci. U.S.*, **55:** 1190–1195.

SKALKA, A. 1968. Personal communication.

SLY, W. S., H. A. EISEN, and L. SIMINOVITCH. 1968. Host survival following infection with or induction of bacteriophage lambda mutants. *Virology*, **34:** 112–127.

SMITH, DAVID H. 1966. Salmonella with transferable drug resistance. *New England Journal of Medicine*, **275:** 626–630.

SMITH, DAVID H. 1967a. R factors mediate resistance to mercury, nickel and cobalt. *Science*, **156:** 1114–1116.

SMITH, DAVID H. 1967b. R-factor-mediated resistance to new aminoglycoside antibiotics. *Lancet*, 252–254.

SMITH, HAMILTON O. 1968. Defective phage formation by lysogens of integration-deficient phage P22 mutants. *Virology*, **34:** 203–223.

SMITH, H. O., and M. LEVINE. 1964. Two sequential repressions of DNA synthesis in the establishment of lysogeny by phage P22 and its mutants. *Proc. Natl. Acad. Sci. U.S.*, **52:** 356–363.

SMITH, HAMILTON O., and M. LEVINE. 1965. Gene order in prophage P22. *Virology*, **27:** 229–231.

SMITH, H. O., and M. LEVINE. 1967. A phage P22 gene controlling integration of prophage. *Virology*, **31:** 207–216.

SOLLER, ARTHUR, LAWRENCE LEVINE, and H. T. EPSTEIN. 1965. The antigenic structure of lambda bacteriophage. *Virology*, **26:** 708–714.

STRACK, H. B., and A. D. KAISER. 1965. On the structure of the ends of λ DNA. *J. Mol. Biol.*, **12:** 36–49.

SUNSHINE, M. B., and B. KELLY. 1967. Studies on P2 prophage-host relationships. I. Alteration of P2 prophage localization patterns in *Escherichia coli*. Transductional heterogenotes associated with the presence of P2 prophage. *Virology*, **32:** 644–653.

SZYBALSKI, W., H. KUBINSKI, *and* P. SHELDRICK. 1966. Pyrimidine clusters on the transducing strand of DNA and their possible role in the initiation of RNA synthesis. *Cold Spring Harbor Symposia Quant. Biol.*, **31:** 123–127.

TAKAHASHI, I. 1965. Localization of spore markers on the chromosome of *Bacillus subtilis*. *J. Bacteriol.*, **89:** 1065–1067.

TAYLOR, A. L. 1963. Bacteriophage-induced mutation in *Escherichia coli*. *Proc. Natl. Acad. Sci. U.S.*, **50:** 1043–1051.

TAYLOR, AUSTIN L., and MARIANNE S. THOMAN. 1964. The genetic map of *Escherichia coli K-12*. *Genetics*, **50:** 659–677.

TAYLOR, AUSTIN L., and CAROL DUNHAM TROTTER. 1967. Revised linkage map of *Escherichia coli*. *Bacteriol. Rev.*, **31:** 332–353.

TAYLOR, MILTON, and CHARLES YANOFSKY. 1966. An explanation for the reduced frequency of double lysogenization. *Virology*, **29:** 502–503.

THOMAS, C. 1968. Personal communication.

THOMAS, R. 1964. On the structure of the genetic segment controlling immunity in temperate bacteriophages. *J. Mol. Biol.*, **8:** 247–253.

THOMAS, RENE. 1966. Control of development in temperate bacteriophages. I. Induction of prophage genes following hetero-immune super-infection. *J. Mol. Biol.*, **22:** 79–95.

THOMAS, R., and L. E. BERTANI. 1964. On the control of the replication of temperate bacteriophages superinfecting immune hosts. *Virology*, **24:** 241–253.

TOMIZAWA, JUN-ICHI, and TOMOKO OGAWA. 1967. Effect of ultraviolet irradiation on bacteriophage lambda immunity. *J. Mol. Biol.*, **23:** 247–263.

UETAKE, HISAO, and T. UCHIDA. 1959. Mutants of *Salmonella* phage ϵ^{15} with abnormal conversion properties. *Virology*, **9:** 495–505.

VOGT, PETER K. 1967. Nonproducing state of Rous Sarcoma cells: its contagiousness in chicken cell cultures. *J. Virology*, **1:** 729–737.

WATANABE, T. 1963. Infective heredity of multiple drug resistance in bacteria. *Bacteriol. Rev.*, **27:** 87–113.

WATANABE, T., and T. FUKASAWA. 1961. Episome-mediated transfer of drug resistance in *Enterobacteriaceae*. III. Tranduction of resistance factors. *J. Bacteriol.*, **82:** 202–208.

WATANABE, TSUTOMU, TOSHIYA TAKANO, TOSHIHIKO ARAI, HIROSHI NISHIDA, and SACHIKO SATO. 1966. Episome-mediated transfer of drug resistance in *Enterobacteriaceae*. X. Restriction and modification of phages by fi^- R factors. *J. Bacteriol.*, **92:** 477–486.

WEIGLE, J. 1966. Assembly of phage lambda in vitro. *Proc. Natl. Acad. Sci. U.S.*, **55:** 1462–1466.

WEISBERG, ROBERT A., and JONATHAN A. GALLANT. 1966. Two functions under *cI* control in lambda lysogens. *Cold Spring Harbor Symposia Quant. Biol.*, **31:** 374–375.

YOUNG, BOBBY G., YOSHIMURA FUKASAWA, and PHILIP E. HARTMAN. 1964. A P22 bacteriophage mutant defective in antigen conversion. *Virology*, **23:** 279–283.

YOUNG, B. G., and PHILIP E. HARTMAN. 1966. Sites of P22 and P221 prophage integration in *Salmonella typhimurium*. *Virology*, **28:** 265–270.

YOUNG, ELTON T., and ROBERT L. SINSHEIMER. 1967. Vegetative bacteriophage DNA. I. Infectivity in a spheroplast assay. *J. Mol. Biol.*, **30:** 147–164.

ZAVADA, V., and E. CALEF. 1968. Integration of $\lambda b2$ in *Escherichia coli* K12-B hybrid. *Virology*, **34:** 128–133.

ZINDER, N. 1958. Lysogenization and superinfection immunity in *Salmonella*. *Virology*, **5:** 291–326.

ZISSLER, JAMES F. 1967. A study of lysogenization by bacteriophage lambda: a phage gene for attachment of prophage. Thesis, University of Rochester.

ZISSLER, JAMES, and ALLAN CAMPBELL. 1968. Manuscript in preparation.

RECENT BOOKS AND REVIEW ARTICLES

DOVE, WILLIAM F. 1968. The genetics of the lambdoid phages. *Ann. Rev. Genet.* (in press).

ECHOLS, HARRISON, and ANNE JOYNER. 1968. The temperate phage. In *The Molecular Basis of Virology* (in press).

FALKOW, STANLEY, E. M. JOHNSON, and L. S. BARON. 1967. Bacterial conjugation and extrachromosomal elements. *Ann. Rev. Genet.*, **1:** 87.

LURIA, S. E., and J. DARNELL. 1967. *General Virology.* Wiley, New York.

NOMURA, M. 1967. Colicins and related bacteriocins. *Ann. Rev. Microbiol.*, **21:** 257–284.

SCAIFE, JOHN. 1967. Episomes. *Ann. Rev. Microbiol.*, **21:** 601–638.

SIGNER, E. 1968. Lysogeny: the integration problem. *Ann. Rev. Microbiol.* (in press).

WATANABE, TSUTOMU. 1967. Evolutionary relationships of R factors with other episomes and plasmids. *Fed. Proc.*, **26:** 23–28.

INDEX OF NAMES

Abe, A., 146
Adelberg, Edward A., 58, 60
Adhya, Sankar, 74, 101, 102, 115
Akiba, T., 40
Alfoldi, L., 11–12
Anderson, E. S., 46–51, 55 n.
Anderson, Thomas F., 29

Baldwin, R. L., 151
Beckwith, J. R., 10, 78–79, 91, 93
Benzer, S., 74
Berg, C. M., 105, 146
Bertani, G., 26, 28, 82
Bertani, L. E., 131–133, 141, 144
Bode, Vernon C., 19, 129
Bonhoeffer, Friedrich, 147
Bordet, J., 25–26
Bott, K. F., 34, 166
Boy de la Tour, E., 21
Brink, R. Alexander, 166
Brinton, Charles C., 36
Bronson, M., 28
Brooks, Katherine, 21, 82, 93
Burgi, E., 149

Cairns, John, 139
Calef, Enrico, 69–71, 73, 110–112, 158–159
Calendar, R., 33 n.
Callan, H. G., 171
Campbell, Allan, 19, 21, 23, 60, 64, 68, 71, 86, 89, 92 n., 96, 101, 104, 146, 152, 156, 158, 160
Caro, Lucien G., 21, 146

Cavalli, Luigi, 7
Chai, N., 143
Chandler, Barbara, 149
Citarella, R. V., 7, 36 n.
Clark, A. J., 63 n., 93
Cleary, P. P., 100
Clowes, R. C., 56, 58
Curtiss, Roy, III, 10, 58, 60, 64, 79, 105
Cuzin, François, 36, 143, 145, 147

Datta, Naomi, 44, 52, 137
Davidoff-Abelson, R., 166
Dawson, G. W. P., 168
Del Campillo Campbell, A., 19, 96, 100
Dove, William F., 19, 142, 148
Dreyer, W. J., 172
Dubnau, E., 43

Echols, Harrison, 92–93, 126
Eisen, Harvey A., 18, 152
Eiserling, F. A., 20, 21, 26
Englesberg, Ellis, 118

Falkow, Stanley, 7, 35, 36 n., 42, 45
Fischer-Fantuzzi, L., 61, 110–112, 158
Franklin, N. C., 71, 73, 76, 113, 141
Fredericq, Pierre, 11, 56
Freifelder, D., 55 n.
Fuerst, C. R., 16
Fukasawa, T., 41, 43, 45–46

Gallant, Jonathan A., 97
Ganesan, A. T., 143

Garen, A., 7–8
Gellert, M., 151
Gilbert, Walter, 117
Gingery, R., 92, 93
Gottesman, M. E., 77, 92
Gough, M., 74
Gratia, J. P., 11, 73, 113
Greenblatt, Irwin M., 166
Gross, Julian D., 10

Hanafusa, H., 168
Harada, Kenji, 43
Harris, A. W., 19
Hartman, Philip E., 30–31
Haruna, L., 142
Hashimoto, Hajime, 42
Hayes, William, 6, 7, 136
Helinski, Donald R., 54
Hershey, A. D., 149
Hirota, Y., 36, 41–42, 62
Hoffman, Donald B., Jr., 82
Hogness, D. S., 23, 25
Hopkins, Nancy, 130
Horiuchi, Tadao, 60, 64
Howard-Flanders, Paul, 100
Hsu, Yu-chih, 31

Ikeda, H., 109
Inman, Ross B., 25
Ippen, Karin, 121
Israel, J. V., 30

Jacob, F., 3, 7–8, 11–13, 16, 36, 47, 58, 60, 66, 69–71, 116–122, 130, 140, 143, 145, 147, 166, 170
Jordan, Elke, 22, 84
Joyner, Anne, 18, 21

Kahn, Phyllis L., 54
Kaiser, A. D., 16, 19, 23, 84 n., 129, 150
Kayajanian, Gary, 23, 101, 102, 104, 142
Kellenberger, Grete, 84
Kelly, B. L., 3, 28
Killen, Karen, 96, 152
Koch, M. A., 169
Kubinski, H., 22

Lark, Karl G., 143, 146
Lederberg, Esther M., 6, 15–16, 37

Lederberg, Joshua, 6, 29, 60, 64, 108–110, 143
Lennox, E., 28
Levine, M., 30–31, 74, 92, 94
Lewis, M. J., 46, 55 n.
Lieb, Margaret, 134
Lindahl, G., 28, 33 n.
Low, Brooks, 10
Luria, S. E., 108
Lwoff, André, 1, 34, 161

Maas, Renata, 145
Maas, Werner K., 63 n.
Mandel, M., 33 n.
Marchelli, C., 111
Matney, Thomas S., 10
McClintock, B., 166–167
Meselson, Matthew, 21–22
Meynell, Elinor, 43, 137
Mitsuhashi, Susumu, 52
Monod, Jacques, 116–122
Moody, E. E. M., 58
Moore, L., 93
Müller-Hill, Benno, 117

Nagata, Toshio, 146
Nagel de Zwaig, R., 54
Nakaya, Rintaro, 40
Nomura, M., 54
Novick, Aaron, 64, 66

Ogawa, Tomoko, 134
Okubo, S., 109

Packman, S., 130
Parkinson, J. S., 18
Pearce, L. E., 43
Pekhov, A. P., 105
Pereira da Silva, L. H., 148
Pittard, James, 58, 80
Ptashne, M., 95, 126, 130, 147

Radding, Charles M., 18, 131
Renaux, E., 25–26
Renshaw, Janet, 58
Richmond, M. H., 44
Ris, Hans, 149
Rogolsky, M., 166
Rolfe, B., 100
Rothman, June L., 72–73, 75–76

INDEX OF NAMES

Rous, Peyton, 168
Rownd, R., 44, 45
Rubenstein, Irwin, 82

Sabin, A. B., 169
Scaife, John, 63 n., 105
Schaeffer, P., 166
Scott, J. R., 27, 33 n.
Shapiro, J., 74, 101
Shaw, W. V., 44
Sheldrick, P., 22
Sheppard, David, 118
Signer, E. R., 3, 10, 19, 74, 87, 91, 93, 96, 159
Siminovitch, L., 17
Sinsheimer, Robert L., 22 n.
Six, E. W., 29, 77–78, 88–89, 91
Skaar, Palmer, 7–8
Skalka, A., 21, 25
Slepecky, R., 166
Sly, W. S., 130
Smith, D. H., 41, 51, 52
Smith, H. O., 30, 31, 74, 77, 92, 94, 96
Smith-Keary, P. F., 168
Soller, Arthur, 21
Spiegelman, S., 142
Stahl, Frank, 149
Stocker, B. A. D., 43
Strack, H. B., 150
Strauss, B., 34
Sunshine, M. B., 3
Szybalski, W., 21, 22, 25

Takahashi, I., 166
Tatum, Edward L., 6
Taylor, A. L., 10, 63 n., 65, 77, 79
Taylor, Milton, 91
Thomas, C., 30, 33 n.
Thomas, Marianne S., 10
Thomas, R., 124, 127, 128, 133, 141, 144
Tomizawa, J., 109–110, 134, 146
Trotter, C. D., 10, 63 n., 65

Uchida, Hisao, 62
Uchida, T., 114
Uetake, Hisao, 114

Vogt, Peter K., 169

Watanabe, T., 40–41, 43, 45–46, 52
Weigle, J., 18, 21–22
Weisberg, Robert A., 97
Wollman, E. L., 3, 7–8, 11–13, 16, 47, 58, 60, 69–71, 130, 166

Yanofsky, Charles, 91
Yarmolinsky, M. B., 77, 92
Young, Bobby C., 30, 31
Young, Elton T., 22 n.

Zavada, V., 73
Zichichi, Maria, 84
Zinder, N. D., 29–31, 108–110
Zissler, James F., 89, 92 n.

SUBJECT INDEX

Acridine dyes, curing of plasmids by, 65; effect on sporulation, 166
Apo-repressors, 117
AraC gene, 118–119
Att (insertion locus), 18, 82
Attachment (chromosomal), of F factors, 8–10; mechanism of, 81–94; modes of, 68–80; of R agents, 43–44; relation to immunity, 135–136; state of, and mode of replication, 147–148
Autonomous replication (independent multiplication), 139–148; of colicinogeny determinant, 11–13; of F factors, 9–10, 38–39; operational criteria for, 57; of phage, 4; relation to chromosomal attachment, 82

Bacterial mating, and chromosome replication, 146–147, 152–153
Bacteriocins, 11
Bacteriophages, *see* Phages

cI gene, 19, 126–127
Cellular differentiation, role of episomes in, 165
Chromosome puffs, 171
Chromosomes (bacterial), 57; attachment of episomes to, *see* Attachment; detachment of episomes from, *see* Detachment; haploidization of, 61; lysogenic, cutting of, 152–153; physical connection with F factors, 143; relation to episomes, 161

Clear plaque mutants, 18, 124–126
Colicinogens, nontransferring, 56
Colicinogeny, 11–13
Colicinogeny determinants (col), 11, 53–56; factors influencing transfer of, 136
Colicins, 11, 53–56
Coliphages, *see* Phages
Composite transfer agents, 38–39
Conjugation (bacterial), agents of, *see* Transfer agents; role of pili in, 136–138
Controlling elements, 166–168
Crosses (bacterial), aberrant inheritance in, 57–59; in genetic mapping of prophage, 2–4
Crosses (interspecific), in study of lambda genetics, 19
Crossover region, 82–87
Cryptic prophage, 110–112, 152
Cryptogenic phage, 110–112
Curing, of lysogenic bacteria, 96–97; of plasmids, 65

Defective mutants, in study of phage genetics, 16–19
Deletion mutants, 65, 73, 84, 104
Δ (delta) factor, 46–51, 55
Detachment (chromosomal), abnormal, 105–108, 112–113; mechanism of, 95–98
Diploids (partial), 60–65; lysogenization of, 92

SUBJECT INDEX

DNA, and gene function, 116; of F agents, 7, 35–36; of lambda, 21–22; modification by P1, 27–28; similarities in phage and bacterium, 87–88; specificity of, 170; synthesis of, 18; in transducing particles, 109–110; transfer in mating, 147

DNA molecules, circularity of, 149–151; hybrid, 108; identified with genetic elements, 38; of infecting particles, 81–82

Double lysogeny, 154–160

Drug resistance, 40, 52; see also Resistance (R) agents

Effector molecules, 117

End joining, 149–153

Enzyme(s), in chromosomal mechanics, 172; in drug resistance, 44; genes affecting synthesis of, 118; "nicking," 151–152; in replication, 142

Epidemiology of R agents, 51–53

Episomes, 12–14, 161–164; and chromosomal mechanics, 170–172; and development, 165–168; gene pickup by, 99–108; interference between, 144–146; segregation at cell division, 143–144

Escherichia coli, Lisbonne-Carrere strain, lysogeny in, 25–26; partially diploid strains, 60–65, 85–87; production of colicins by, 11; in transfer of drug resistance, 40–42, 46

E. coli K-12, fertility agent of, 6–10, 35–37; isolation of lambda from, 15–16; location of prophages on, 3; in studies of attachment sites, 85–87

F′ factors, 105–108

Fertility (F) agents (factors), 6–10, 35–39, 55; attachment to chromosomes, 8–10, 78–80; chromosomal splitting by, 152–153; detachment of, 97–98; integration of, 94; interference with Hfr, 9, 144–146; physical connection to chromosome, 143; transducing variants of, 105–108; transfer of, 136–138

Fertility inhibition (fi) characters in R agents, 46, 137–138

Fine structure genetics, 38

Gene function, 116–117; inactivation at insertion site, 79–80; relation to attachment, 135–136

Gene order, in double lysogens, 154–159; in prophage and free phage, 69–70

Gene pickup, 37, 99–108

Gene regulation, 115–122

Generalized transduction, 108–110

Genes, assignment to operons, 122; chromosomal, transfer of, 7–8, 58–59; mating, 41; regulator, 117

Genetic elements, identification with nucleic acid molecules, 38

Haploidization, of diploids, 61

Heat-sensitive mutants, 17, 66

Helper phage, 22

Heterochromatin, as mutator, 167

Heteroimmune phages, 122–124

Hfr character, 7–10; interference with F factor, 137–138, 144–146; in transfer of chromosomal genes, 58

Homoimmune phages, 122

Immunity (superinfection), effect on replication, 141–142; genetic control of, 122–131; and repression, 114–115; relation to attachment, 135–136; relation to prophage localization, 83–84; and steric facilitation, 89–92; of transfer factors, 136–138; titration of, 131–135

Immunity repressor, *see* Repressor

Independent multiplication, *see* Autonomous replication

Inducer, 117–118

Induction curing, 97

Infection, acquisition of lysogeny by, 1; *see also* Superinfection

Inheritance, aberrant, in crosses, 57–59; nonchromosomal, 66–67

Initiation site, 139–140

Initiator, 140

Insertion, 61–71; genes controlling, 18; and inactivation of bacterial genes,

Insertion (*continued*)
79–86; model for, 72–77; physical nature of, 81–82; requirements for, 149; specificity of, 19
Integrase, mutants, 92–94; in double lysogens, 159; and attachment, 91–92; and detachment, 95–96
Integrated state, 4
Integration, 69, 85, 92–94
Interference, between autonomous and integrated states, 9; between episomes, 144–146

Lambda (phage), *see* Phage lambda
Linkage analysis, of transfer agents, 39
Lysogenization, frequency of, 135–136; genes regulating, 18–19; mechanism of, 82; of partial diploids, 92
Lysogeny, 1–5; and colicinogeny, 54; as model for viral induction of tumors, 169; multiple, 154–160; physiological problems posed by, 114–115

Mating gene (*m*) of R agents, 41
Metal ions, plasmid curing by, 65
Mitochondrion, genetic aspects of, 143–144
Mobilization, 58–59
Multiplication, *see* Replication
Multiplicity effect, 131–135
Mutants (mutations), defective, 16–19; deletion, 65, 73–74, 84, 104–105; heat-sensitive, 17–18, 66; involving integrase, 92–94; operator, 129–131; replication-deficient, 66; regulator, 97, 124–126; suppressor-sensitive, 17–18; transfer-deficient, 42–43; virulent, 130
Mutator phage (mu), 79–80

Natural selection, in episomes, 162–163
Nicking enzyme, 151–152, 159
Nucleic acid, *see* DNA

Operator, location of, 120–121
Operator mutants, 129–131
Operator regions, 117, 129
Operons, 116–117

Phage(s), attachment sites of, 82–83; cryptogenic, 110–112; defective, 110; genes controlling formation of, 17–18; helper, 22; temperate, 2, 15–34; transducing, 24–25, 74, 98, 104–105; virulent, 2
Phage 18, 3, 76–77
Phage 21, 3, 83
Phage 80, 3, 73–74, 83–84
Phage 363, 28
Phage 434, 3, 83–84
"Phage carrier" state, 31
Phage formation, multiplicity dependence of, 133
Phage lambda, 15–25, 32–33; attachment site of, 83; in attachment studies, 68–80; DNA of, 21–22, 87; genetic program of, 126–129; immunity region of, 125; integration of, 69; production of factors affecting phage growth, 141–142; recombinase in, 119–120; recombination of, 70, 71; regulator mutants of, 124–126; specificity of attachment of, 77; spontaneous annealing of, 150; steric hindrance and facilitation in, 91–92; topological mapping of, 101, 102; transducing variants of, 105–106; virulent mutations of, 130
Phage lambda *b2*, 84–87
Phage lambda *c*-17, 130
Phage lambda *crg*, 110–112
Phage lambda *db*, 142
Phage lambda *dg*, 99–104, 158–159
Phage lambda *ind*, 126
Phage lambda *int*, 92–94, 159
Phage mu, 79–80
Phage P1, 25–28, 32–33, 41–42, 108–110
Phage P2, 28–29, 32–33, 77–78, 83–84, 89, 91
Phage P22 (*Salmonella*), 29–33, 41–43
Phage SP10, 31–32
Pili, 36, 136–138
Plasmids, 57; curing of, 65; distinguished from chromosomes, 161; loss of, 59; R agents as, 47–50, staphylococcal, 66–67
Polar effect, 122

Polylysogeny, 154–160
Principle of progressive rarity, 156
Promoter, 120–121
Prophage, 1; attachment of, *see* Attachment; chromosomal location of, 2–4; cryptic, 110–112; insertion of, *see* Insertion; integration of, 69, 85, 92–94; localization of, and immunity, 83–84; loss of, 97, 154–159; recombination of, 69; splitting of chromosome by, 152; in bacterial spores, 31–34.
Proteins, in colicins, 54; phage-specific, 18; recognition of nucleotide sequences by, 93
Puffs, in chromosomes, 171

R222, 40–45, 55
Radiation curing, 96
Recognition region, 120–121
Recombination, during attachment, 81–82; of lambda prophages, 70, 71
Recombinase, 90–91, 93, 119–120, 159
Recombinase system, bacterial, 58; in lambda (*red*), 18
Regulator gene, 117, 118
Regulator mutants, 18, 124–126
Replication, autonomous, *see* Autonomous replication; control of, 139–140; immunity-specific, 141–142; in integrated state, 146–147; mode of, 147–148
Replication-deficient mutants, 66
Replicator, 140
Replicon, 140
Repression, 114–115; control of, 18–19; by R factors, 136–138; *see also* Immunity
Repressor, 117; control of phage replication by, 141; saturation of, 131–135
Resistance (R) agents (factors), 39–53; repression by, 136–138; role of Δ factor in, 49–50
Resistance transfer factors (RTF), 39–41

RNA, messenger, 116; Qβ, replication of, 142
Rous Sarcoma virus, 168–169

Salmonella, controlling elements in, 168; monitoring of, 51; in transfer of drug resistance, 41–42, 46
Segregation, in doubly lysogenic bacteria, 71–72; of episomes at cell division, 143–144; vegetative, 59–65
Shigella, action of phages on, 26–29; drug resistance in, 40; *flexnerii* 2b strain 222, R agent of, 40–45; transduction effects in, 108
Sporulation, 165–166
Staphylococcus, nonchromosomal inheritance in, 66–67
Steric facilitation, 89–92
Steric hindrance, 88–89
Superinfection, in double lysogens, 157–158; of heteroimmune lysogens, 128; multiplicity effects in, 132–135; study of prophage genetics by, 4
Superinfection curing, 97
Suppressor-sensitive mutants, 17–18
Synthesis, *see* Replication

Transcription, 116, 119, 126
Transducing phages, 24–25, 74, 98, 105
Transduction, generalized, 108–110
Transfer agents (factors), 35–36; composite, 38–39; and drug resistance, 47; immunity of, 136–138; mobilization by, 58–59; properties of, 55
Transfer-deficient mutants, 42–43
Tumor viruses, 168–170

Vegetative multiplication, 4
Vegetative segregation, 59–65
Viruses, definition of, 161–162; operation of natural selection on, 162–163; tumor, 168–170; *see also* Phage(s)

70 71   7 6 5 4 3 2